The Art of Designing Embedded Systems

The Art of Designing Embedded Systems

Jack G. Ganssle

Newnes

BOSTON OXFORD AUCKLAND JOHANNESBURG MELBOURNE NEW DELHI

Newnes is an imprint of Butterworth–Heinemann.

Copyright © 2000 by Butterworth–Heinemann

⦵ A member of the Reed Elsevier group

All rights reserved.

No part of this publication may be reproduced, stored in a retrieval system, or transmitted in any form or by any means, electronic, mechanical, photocopying, recording, or otherwise, without the prior written permission of the publisher.

∞ Recognizing the importance of preserving what has been written, Butterworth–Heinemann prints its books on acid-free paper whenever possible.

 Butterworth–Heinemann supports the efforts of American Forests and the Global ReLeaf program in its campaign for the betterment of trees, forests, and our environment.

Library of Congress Cataloging-in-Publication Data

Ganssle, Jack G.
    The art of designing embedded systems / Jack G. Ganssle.
      p.   cm.
    ISBN 0-7506-9869-1 (hc. : alk. paper)
    1. Embedded computer systems—Design. I. Title.
Tk7895.E42G36   1999   99-36724
004.16—dc21   CIP

British Library Cataloguing-in-Publication Data
A catalogue record for this book is available from the British Library.

The publisher offers special discounts on bulk orders of this book.
For information, please contact:
Manager of Special Sales
Butterworth-Heinemann
225 Wildwood Avenue
Woburn, MA 01801-2041
Tel: 781-904-2500
Fax: 781-904-2620

For information on all Butterworth–Heinemann publications available, contact our World Wide Web home page at: http://www.newnespress.com

10 9 8 7 6 5 4 3 2

Printed in the United States of America

Dedicated to Graham and Kristy

Contents

Acknowledgments *ix*

Chapter 1 Introduction *1*

Chapter 2 Disciplined Development *5*

Chapter 3 Stop Writing Big Programs! *35*

Chapter 4 Real Time Means Right Now *53*

Chapter 5 Firmware Musings *87*

Chapter 6 Hardware Musings *109*

Chapter 7 Troubleshooting Tools *133*

Chapter 8 Troubleshooting *165*

Chapter 9 People Musings *187*

Appendix A A Firmware Standards Manual *203*

Appendix B A Drawing System *223*

Index *237*

Acknowledgments

I'd like to thank Pam Chester, my editor at Butterworth–Heinemann, for her patience and good humor through the birthing of this book. And thanks to Joe Beitzinger for his valuable comments on the initial form of the book.

Finally, thanks to the many developers I've worked with over the years, and the many more who have corresponded.

CHAPTER **1**

Introduction

Any idiot can write code. Even teenagers can sling gates and PAL equations around. What is it that separates us from these amateurs? Do years of college necessarily make us professionals, or is there some other factor that clearly delineates *engineers* from *hackers*? With the phrase "sanitation engineer" now rooted in our lexicon, is the real meaning behind the word *engineer* cheapened?

Other professions don't suffer from such casual word abuse. Doctors and lawyers have strong organizations that, for better or worse, have changed the law of the land to keep the amateurs out. You just don't find a teenager practicing medicine, so "doctor" conveys a precise, strong meaning to everyone.

Lest we forget, the 1800s were known as "the great age of the engineer." Engineers were viewed as the celebrities of the age, as the architects of tomorrow, the great hope for civilization. (For a wonderful description of these times, read *Isamard Kingdom Brunel*, by L.T.C. Rolt.)

How things have changed!

Our successes at transforming the world brought stink and smog, factories weeping poisons, and landfills overflowing with products made obsolete in the course of months. The *Challenger* explosion destroyed many people's faith in complex technology (which shows just how little understanding Americans have of complexity). An odd resurgence of the worship of the primitive is directly at odds with the profession we embrace. Declining test scores and an urge to make a lot of money *now* means that U.S. engineering enrollments have declined 25% in the decade from 1988 to 1997.

All in all, as Rodney Dangerfield says, "We just can't get no respect."

It's my belief that this attitude stems from a fundamental misunderstanding of what an engineer is. We're not scientists, trying to gain a new understanding of the nature of the universe. Engineers are the world's problem solvers. We convert dreams to reality. We bridge the gap between pure researchers and consumers.

Problem solving is surely a noble profession, something of importance and fundamental to the future viability of a complex society. Suppose our leaders were as single-mindedly dedicated to problem solving as is any engineer: we'd have effective schools, low taxation, and cities of light and growth rather than decay. Perhaps too many of us engineers lack the social nuances to effectively orchestrate political change, but there's no doubt that our training in problem solving is ultimately the only hope for dealing with the ecological, financial, and political crises coming in the next generation.

My background is in the embedded tool business. For two decades I designed, built, sold, and supported development tools, working with thousands of companies, all of whom were struggling to get an embedded product out the door, on time and on budget. Few succeed. In almost all cases, when the widget was finally complete (more or less; maintenance seems to go on forever because of poor quality), months or even years late, the engineers took maybe five seconds to catch their breath and then started on yet another project. Rare was the individual who, after a year on a project, sat and thought about what went right and wrong on the project. Even rarer were the people who engaged in any sort of process improvement, of learning new engineering techniques and applying them to their efforts. Sure, everyone learns new tools (say, for ASIC and FPGA design), but few understood that *it's just as important to build an effective way to design products, as it is to build the product.* We're not applying our problem-solving skills to the way we work.

In the tool business I discovered a surprising fact: most embedded developers work more or less in isolation. They may be loners designing all of the products for a company, or members of a company's design team. The loner and the team are removed from others in the industry, so they develop their own generally dysfunctional habits that go forever uncorrected. Few developers or teams ever participate in industry-wide events or communicate with the rest of the industry. We, who invented the communications age, seem to be incapable of using it!

One effect of this isolation is a hardening of the development arteries: we are unable to benefit from others' experiences, so we work ever harder without getting smarter. Another is a feeling of frustration, of thinking, "What is wrong with us—why are our projects so much more a problem than anyone else's?" In fact, most embedded developers are in the same boat.

This book comes from seeing how we all share the same problems while not finding solutions. Never forget that engineering is about solving problems . . . including the ones that plague the way we engineer!

Engineering is the process of making choices; make sure yours reflect simplicity, common sense, and a structure with growth, elegance, and flexibility, with debugging opportunities built in.

In general, we all share these same traits and the inescapable problems that arise from them:

- We jump from design to building too fast. Whether it's writing code or drawing circuits, the temptation to be doing rather than thinking inevitably creates disaster.
- We abdicate our responsibility to be part of the project's management. When we blindly accept a feature set from marketing we're inviting chaos: only engineering can provide a rational cost/benefit tradeoff. Acceding to capricious schedules figuring that heroics will save the day is simply wrong. When we're not the boss, then we simply must manage the boss: educate, cajole, and demonstrate the correct ways to do things.
- We ignore the advances made in the past 50 years of software engineering. Most teams write code the way they did at age 15, when better ways are well known and *proven*.
- We accept lousy tools for lousy reasons. In this age of leases, loans, and easy money, there's always a way to get the stuff we need to be productive. Usually a nattily attired accountant is the procurement barrier, a rather stunning development when one realizes that the accountant's role is not to stop spending, but to spend in a cost-effective manner. The basic lesson of the industrial revolution is that capital investment is a critical part of corporate success.
- And finally, a theme I see repeated constantly is that of poor detail management. Projects run late because people forget to do simple things. Never have we had more detail management tools, from PDAs to personal assistants to conventional Daytimers and Day Runners. One afternoon almost a decade ago I looked up from a desk piled high with scraps of paper listing phone calls and to-dos and let loose a primal scream. At the time I went on a rampage,

looking for some system to get my life organized so I knew *what* to do *when*. For me, an electronic Daytimer—coupled with a determination to use it every hour of every day—works. The first thing that happens in the morning is the organizer pops up on my screen, there to live all day long, checked and updated constantly. Now I never (well, *almost* never) forget meetings or things I've promised to do.

And so, I see a healthy engineering environment as the right mix of technology, skills, and processes, all constantly evaluated and managed.

CHAPTER **2**

Disciplined Development

Software engineering is not a discipline. Its practitioners cannot systematically make and fulfill promises to deliver software systems on time and fairly priced.

—Peter Denning

The seduction of the keyboard is the downfall of all too many embedded projects.

Writing code is fun. It's satisfying. We feel we're making progress on the project. Our bosses, all too often unskilled in the nuances of building firmware, look on approvingly, smiling that we're clearly accomplishing something worthwhile.

As a young developer working on assembly-language-based systems, I learned to expect long debugging sessions. Crank some code, and figure on months making it work. Debugging is hard work (but fun—it's great to play with the equipment all the time!), so I learned to budget 50% of the project time to chasing down problems.

Years later, while making and selling emulators, I saw this pattern repeated, constantly, in virtually every company I worked with. In fact, this very approach to building firmware is a godsend to the tool companies who all thrive on developers' poor practices and resulting sea of bugs. Without bugs, debugger vendors would be peddling pencils.

A quarter century after my own first dysfunctional development projects, in my travels lecturing to embedded designers, I find the pattern remains unbroken. The rush to write code overwhelms all common sense.

The overused word "process" (note that only the word is overused; the concept itself is sadly neglected in the firmware world) has garnered enough attention that some developers claim to have institutionalized a reasonable way to create software. Under close questioning, though, the majority of these admit to applying their rules in a haphazard manner.

When the pressure heats up—the very time when sticking to a system that *works* is most needed—most succumb to the temptation to drop the systems and just crank out code.

> As you're boarding a plane you overhear the pilot tell his right-seater, "We're a bit late today; let's skip the take-off checklist." Absurd? Sure. Yet this is precisely the tack we take as soon as deadlines loom; we abandon all discipline in a misguided attempt to beat our code into submission.

Any Idiot Can Write Code

In their studies of programmer productivity, Tom DeMarco and Tim Lister found that all things being equal, programmers with a mere 6 months of experience typically perform as well as those with a year, a decade, or more.

As we developers age we get more experience—but usually the same experience, repeated time after time. As our careers progress we justify our escalating salaries by our perceived increasing wisdom and effectiveness. Yet the data suggests that the *value of experience is a myth*.

Unless we're prepared to find new and better ways to create firmware, and until we implement these improved methods, we're no more than a step above the wild-eyed teen-aged guru who lives on Coke and Twinkies while churning out astonishing amounts of code.

Any idiot can create code; professionals find ways to consistently create high-quality software on time and on budget.

Firmware Is the Most Expensive Thing in the Universe

Norman Augustine, former CEO of Lockheed Martin, tells a revealing story about a problem encountered by the defense community. A high-performance fighter aircraft is a delicate balance of conflicting needs: fuel range versus performance. Speed versus weight. It seemed that by the late 1970s fighters were at about as heavy as they'd ever be. Contractors, always pursuing larger profits, looked in vain for something they could add that cost a lot, but that weighed nothing.

The answer: firmware. Infinite cost, zero mass. Avionics now accounts for more than 40% of a fighter's cost.

Two decades later nothing has changed ... except that firmware is even more expensive.

What Does Firmware Cost?

Bell Labs found that to achieve 1–2 defects per 1000 lines of code they produce 150 to 300 lines per month. Depending on salaries and overhead, this equates to a cost of around $25 to $50 per line of code.

Despite a lot of unfair bad press, IBM's space shuttle control software is remarkably error free and may represent the best firmware ever written. The cost? $1000 per statement, for no more than one defect per 10,000 lines.

Little research exists on embedded systems. After asking for a per-line cost of firmware I'm usually met with a blank stare followed by an absurdly low number. "$2 a line, I guess" is common. Yet, a few more questions (How many people? How long from inception to shipping?) reveals numbers an order of magnitude higher.

Anecdotal evidence, crudely adjusted for reality, suggests that if you figure your code costs $5 a line you're lying—or the code is junk. At $100/line you're writing software documented almost to DOD standards. Most embedded projects wind up somewhere in between, in the $20–40/line range. There are a few gurus out there who consistently do produce quality code much cheaper than this, but they're on the 1% asymptote of the bell curve. If you feel you're in that select group—we all do—take data for a year or two. Measure time spent on a project from inception to completion (with all bugs fixed) and divide by the program's size. Apply your loaded salary numbers (usually around twice the number on your paycheck stub). You'll be surprised.

Quality Is Nice ... As Long As It's Free

The cost data just described is correlated to a quality level. Since few embedded folks measure bug rates, it's all but impossible to add the quality measure into the anecdotal costs. But quality does indeed have a cost.

We can't talk about quality without defining it. Our intuitive feel that a bug-free program is a high-quality program is simply wrong. Unless you're using the Netscape "give it away for free and make it up in volume" model, we write firmware for one reason only: profits. Without profits the engineering budget gets trimmed. Without profits the business eventually fails and we're out looking for work.

Happy customers make for successful products and businesses. The customer's delight with our product is the ultimate and only important measure of quality.

Thus: *the quality of a product is exactly what the customer says it is.*

Obvious software bugs surely mean poor quality. A lousy user interface equates to poor quality. If the product doesn't quite serve the buyer's needs, the product is defective.

It matters little whether our code is flaky or marketing overpromised or the product's spec missed the mark. The company is at risk because of a quality problem, so we've all got to take action to cure the problem.

No-fault divorce and no-fault insurance acknowledge the harsh realities of trans-millennium life. We need a no-fault approach to quality as well, to recognize that no matter where the problem came from, we've all got to take action to cure the defects and delight the customer.

This means that when marketing comes in a week before delivery with new requirements, a mature response from engineering is not a stream of obscenities. Maybe . . . just maybe . . . marketing has a point. We make mistakes (and spend heavily on debugging tools to fix them). So does marketing and sales.

Substitute an assessment of the proposed change for curses. Quality is not free. If the product will not satisfy the customer as designed, if it's not till a week before shipment that these truths become evident, then let marketing et al. know the impact on the cost and the schedule.

Funny as the "Dilbert" comic strip is, it does a horrible disservice to the engineering community by reinforcing the hostility between engineers and the rest of the company. The last thing we need is more confrontation, cynicism, and lack of cooperation between departments. We're on a mission: *make the customer happy*! That's the *only* way to consistently drive up our stock options, bonuses, and job security.

Unhappily, "Dilbert" does portray too many companies all too accurately. If your outfit requires heroics all the time, if there's no (polite) communication between departments, then something is broken. Fix it or leave.

The CMM

Few would deny that firmware is a disaster area, with poor-quality products getting to market late and over budget. Don't become resigned to the status quo. As engineers we're paid to solve problems. No problem is greater, no problem is more important, than finding or inventing faster, better ways to create code.

The Software Engineering Institute's (www.sei.cmu.edu) Capability Maturity Model (CMM) defines five levels of software maturity and outlines a plan to move up the scale to higher, more effective levels:

1. *Initial*—Ad hoc and Chaotic. Few processes are defined, and success depends more on individual heroic efforts than on following a process and using a synergistic team effort.
2. *Repeatable*—Intuitive. Basic project management processes are established to track cost, schedule, and functionality. Planning and managing new products are based on experience with similar projects.
3. *Defined*—Standard and Consistent. Processes for management and engineering are documented, standardized, and integrated into a standard software process for the organization. All projects use an approved, tailored version of the organization's standard software process for developing software.
4. *Managed*—Predictable. Detailed software process and product quality metrics establish the quantitative evaluation foundation. Meaningful variations in process performance can be distinguished from random noise, and trends in process and product qualities can be predicted.
5. *Optimizing*—Characterized by Continuous Improvement. The organization has quantitative feedback systems in place to identify process weaknesses and strengthen them proactively. Project teams analyze defects to determine their causes; software processes are evaluated and updated to prevent known types of defects from recurring.

Captain Tom Schorsch of the U.S. Air Force realized that the CMM is just an optimistic subset of the true universe of development models. He discovered the CIMM—Capability Immaturity Model—which adds four levels from 0 to −3:

0. *Negligent*—Indifference. Failure to allow successful development process to succeed. All problems are perceived to be technical problems. Managerial and quality assurance activities are deemed to be overhead and superfluous to the task of software development process.

−1. *Obstructive*—Counterproductive. Counterproductive processes are imposed. Processes are rigidly defined and adherence to the form is stressed. Ritualistic ceremonies abound. Collective management precludes assigning responsibility.

> −2. *Contemptuous*—Arrogance. Disregard for good software engineering institutionalized. Complete schism between software development activities and software process improvement activities. Complete lack of a training program.
>
> −3. *Undermining*—Sabotage. Total neglect of own charter, conscious discrediting of organization's software process improvement efforts. Rewarding failure and poor performance.
>
> If you've been in this business for a while, this extension to the CMM may be a little too accurate to be funny. . . .

The idea behind the CMM is to find a defined way to predictably make good software. The words "predictable" and "consistently" are the keynotes of the CMM. Even the most dysfunctional teams have occasional successes—generally surprising everyone. The key is to change the way we build embedded systems so we are *consistently* successful, and so we can reliably *predict* the code's characteristics (deadlines, bug rates, cost, etc.).

Figure 2-1 shows the result of using the tenants of the CMM in achieving schedule and cost goals. In fact, level 5 organizations don't always deliver on time. The probability of being on time, though, is high and the typical error bands low.

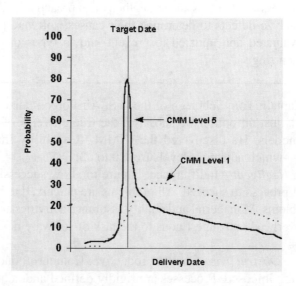

FIGURE 2-1 Improving the process improves the odds of meeting goals and narrows the error bands.

Compare this to the performance of a Level 1 (Initial) team. The odds of success are about the same as at the craps tables in Las Vegas. A 1997 survey in *EE Times* confirms this data in their report that 80% of embedded systems are delivered late.

One study of companies progressing along the rungs of the CMM found the following *per year* results:

- 37% gain in productivity
- 18% more defects found pre-test
- 19% reduction in time to market
- 45% reduction in customer-found defects

It's pretty hard to argue with results like these. Yet the vast majority of organizations are at Level 1 (see Figure 2-2). In my discussions with embedded folks, I've found most are only vaguely aware of the CMM. An obvious moral is to study constantly. Keep up with the state of the art of software development.

Figure 2-2 shows a slow but steady move from Level 1 to 2 and beyond, suggesting that anyone not working on their software processes will be as extinct as the dinosaurs. You cannot afford to maintain the status quo unless your retirement is near.

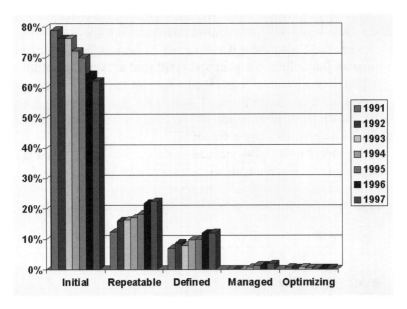

FIGURE 2-2 Over time companies are refining their development processes.

At the risk of being proclaimed a heretic and being burned at the stake of political incorrectness, I advise most companies to be wary of the CMM. Despite its obvious benefits, the pursuit of CMM is a difficult road all too many companies just cannot navigate. Problems include the following:

1. Without deep management commitment CMM is doomed to failure. Since management rarely understands—or even cares about—the issues in creating high-quality software, their tepid buy-in all too often collapses when under fire from looming deadlines.
2. The path from level to level is long and tortuous. Without a passionate technical visionary guiding the way and rallying the troops, individual engineers may lose hope and fall back on their old, dysfunctional software habits.

CMM is a tool. Nothing more. Study it. Pull good ideas from it. Proselytize its virtues to your management. But have a backup plan you can realistically implement *now* to start building better code immediately. Postponing improvement while you "analyze options" or "study the field" always leads back to the status quo. Act now!

Solving problems is a high-visibility process; preventing problems is low-visibility. This is illustrated by an old parable:

In ancient China there was a family of healers, one of whom was known throughout the land and employed as a physician to a great lord. The physician was asked which of his family was the most skillful healer. He replied, "I tend to the sick and dying with drastic and dramatic treatments, and on occasion someone is cured and my name gets out among the lords."

"My elder brother cures sickness when it just begins to take root, and his skills are known among the local peasants and neighbors."

"My eldest brother is able to sense the spirit of sickness and eradicate it before it takes form. His name is unknown outside our home."

The Seven-Step Plan

Arm yourself with one tool—one tool only—and you can make huge improvements in both the quality and delivery time of your next embedded project.

That tool is *an absolute commitment to make some small but basic changes to the way you develop code.*

Given the will to change, here's what you should do *today*:

1. Buy and use a Version Control System.
2. Institute a Firmware Standards Manual.
3. Start a program of Code Inspections.
4. Create a quiet environment conducive to thinking.

More on each of these in a few pages. Any attempt to institute just one or two of these four ingredients will fail. All couple synergistically to transform crappy code to something you'll be proud of.

Once you're up to speed on steps 1–4, add the following:

5. Measure your bug rates.
6. Measure code production rates.
7. Constantly study software engineering.

Does this prescription sound too difficult? I've worked with companies that have implemented steps 1–4 in *one day*! Of course they tuned the process over a course of months. That, though, is the very meaning of the word "process"—something that constantly evolves over time.

But the benefits accrue as soon as you start the process. Let's look at each step in a bit more detail.

Step 1: Buy and Use a VCS

Even a one-person shop needs a formal VCS (Version Control System). It is truly magical to be able to rebuild any version of a set of firmware, even one many years old. The VCS provides a sure way to answer those questions that pepper every bug discussion, such as "When did this bug pop up?"

The VCS is a database hosted on a server. It's the repository of all of the company's code, make files, and the other bits and pieces that make up a project. There's no reason not to include hardware files as well—schematics, artwork, and the like.

A VCS insulates your code from the developers. It keeps people from fiddling with the source; it gives you a way to track each and every change. It controls the number of people working on modules, and provides mechanisms to create a single correct module from one that has been (in error) simultaneously modified by two or more people.

Sure, you can sneak around the VCS, but like cheating on your taxes there's eventually a day of reckoning. Maybe you'll get a few minutes of time savings up front . . . inevitably followed by hours or days of extra time paying for the shortcut.

Never bypass the VCS. Check modules in and out as needed. Don't hoard checked-out modules "in case you need them." Use the system as intended, daily, so there's no VCS cleanup needed at the project's end.

The VCS is also a key part of the file backup plan. In my experience it's foolish to rely on the good intentions of people to back up religiously. Some are passionately devoted; others are concerned but inconsistent. All too often the data is worth more than all of the equipment in a building— even more than the building itself. Sloppy backups spell eventual disaster.

I admit to being anal-retentive about backups. A fire that destroys all of the equipment would be an incredible headache, but a guaranteed business-buster is the one that smokes the data.

Yet, preaching about data duplication and implementing draconian rules is singularly ineffective.

A VCS saves all project files on a single server, in the VCS database. Develop a backup plan that saves the VCS files each and every night. With the VCS there's but one machine whose data is life and death for the company, so the backup problem is localized and tractable. Automate the process as much as possible.

One Saturday morning I came into the office with two small kids in tow. Something seemed odd, but my disbelief masked the nightmare. Awakening from the fog of confusion I realized all of engineering's computers were missing! The entry point was a smashed window in the back. Fearful there was some chance the bandits were still in the facility I rushed the kids next door and called the cops.

The thieves had made off with an expensive haul of brand-new computers, including the server that hosted the VCS and other critical files. The most recent backup tape, which had been plugged into the drive on the server, was also missing.

Our backup strategy, though, included daily tape rotation into a fireproof safe. After delighting the folks at Dell with a large emergency computer order, we installed the one-day-old tape and came back up with virtually no loss of data.

If you have never had an awful, data-destroying event occur, just wait. It will surely happen. Be prepared.

Checkpoint Your Tools

An often overlooked characteristic of embedded systems is their astonishing lifetime. It's not unusual to ship a product for a decade or more. This implies that you've got to be prepared to support old versions of every product.

As time goes on, though, the tool vendors obsolete their compilers, linkers, debuggers, and the like. When you suddenly have to change a product originally built with version 2.0 of the compiler—and now only version 5.3 is available—what are you going to do? The new version brings new risks and dangers. At the very least it will inflict a host of unknowns on your product. Are there new bugs? A new code generator means that the real-time performance of the product will surely differ. Perhaps the compiled code is bigger, so it no longer fits in ROM.

It's better to simply use the original compiler and linker throughout the product's entire lifecycle, so *preserve the tools*. At the end of a project check all of the tools into the VCS. It's cheap insurance.

When I suggested this to a group of engineers at a disk drive company, the audience cheered! Now that big drives cost virtually nothing, there's no reason not to go heavy on the mass storage and save everything.

A lot of vendors provide version control systems. One that's cheap, very intuitive, and highly recommended is Microsoft's SourceSafe.

The frenetic march of technology creates yet another problem we've largely ignored: today's media will be unreadable tomorrow. Save your tools on their distribution CD-ROMs and surely in the not-too-distant future CD-ROMs will be supplanted by some other, better, technology. In time you'll be unable to find a CD-ROM reader.

The VCS lives on your servers, so it migrates with the advance of technology. If you've been in this field for a while, you've tossed out each generation of unreadable media: can you find a drive that will read an 8-inch floppy anymore? How about a 160K 5-inch disk?

Step 2: Institute a Firmware Standards Manual

You can't write good software without a consistent set of code guidelines. Yet, the vast majority of companies have no standards—no written and enforced baseline rules. A commonly cited reason is the lack of such standards in the public domain. So, I've removed this excuse by including a firmware standard in Appendix A.

Not long ago there were so many dialects of German that people in neighboring provinces were quite unable to communicate with each other, though they spoke the same nominal language. Today this problem is manifested in our code. Though the programming languages have international standards, unless we conform to a common way of expressing our ideas within the language, we're coding in personal dialects. Adopt a standard way of writing your firmware, and reject code that strays from the standard.

The standard ensures that all firmware developed at your company meets minimum levels of readability and maintainability. Source code has two equally important functions: it must *work*, and it must clearly *communicate how it works* to a future programmer, or to the future version of yourself. Just as standard English grammar and spelling make prose readable, standardized coding conventions illuminate the software's meaning.

A peril of instituting a firmware standard is the wildly diverse opinions people have about inconsequential things. Indentation is a classic example: developers will fight for months over quite minor issues. The only important thing is to *make a decision*. "We are going to indent in this manner. Period." Codify it in the standard, and then hold all of the developers to those rules.

Step 3: Use Code Inspections

There *is* a silver bullet that can drastically improve the rate at which you develop code while also reducing bugs. Though this bit of magic can reduce debugging time by an easy factor of 10 or more, despite the fact that it's a technique well known since 1976, and even though neither tools nor expensive new resources are needed, few embedded folks use it.

Formal Code Inspections are probably the most important tool you can use to get your code out faster with fewer bugs. The inspection plays on the well-known fact that "two heads are better than one." The goal is to identify and remove bugs *before* testing the code.

Those that are aware of the method often reject it because of the assumed "hassle factor." Usually few developers are aware of the benefits that have been so carefully quantified over time. Let's look at some of the data.

- The very best of inspection practices yield stunning results. For example, IBM manages to remove 82% of all defects before testing even starts!

- One study showed that, as a rule of thumb, each defect identified during inspection saves around 9 hours of time downstream.
- AT&T found inspections led to a 14% increase in productivity and a tenfold increase in quality.
- HP found that 80% of the errors detected during inspections were unlikely to be caught by testing.
- HP, Shell Research, Bell Northern, and AT&T all found inspections 20 to 30 times more efficient than testing in detecting errors.
- IBM found that inspections gave a 23% increase in productivity and a 38% reduction in bugs detected after unit test.

So, though the inspection may cost up to 20% more time up front, debugging can shrink by an order of magnitude or more. The reduced number of bugs in the final product means you'll spend less time in the mind-numbing weariness of maintenance as well.

There is no known better way to find bugs than through Code Inspections! Skipping inspections is a sure sign of the amateur firmware jockey.

The Inspection Team

The best inspections come about from properly organized teams. *Keep management off the team.* Experience indicates that when a manager is involved usually only the most superficial bugs are caught, since no one wishes to show the author to be the cause of major program defects.

Four formal roles exist: the Moderator, Reader, Recorder, and Author.

The Moderator, always technically competent, leads the inspection process. He or she paces the meeting, coaches other team members, deals with scheduling a meeting place and disseminating materials before the meeting, and follows up on rework (if any).

The Reader takes the team through the code by paraphrasing its operation. Never let the Author take this role, since he may read what he meant instead of what was implemented.

A Recorder notes each error on a standard form. This frees the other team members to focus on thinking deeply about the code.

The Author's role is to understand the errors and to illuminate unclear areas. As Code Inspections are never confrontational, the Author should never be in a position of defending the code.

An additional role is that of Trainee. No one seems to have a clear idea how to create embedded developers. One technique is to include new folks (only one or two per team) into the Code Inspection. The Trainee then gets a deep look inside the company's code, and an understanding of how the code operates.

It's tempting to reduce the team size by sharing roles. Bear in mind that Bull HN found four-person inspection teams to be twice as efficient and twice as effective as three-person teams. A Code Inspection with three people (perhaps using the Author as the Recorder) surely beats none at all, but do try to fill each role separately.

The Process

Code Inspections are a *process* consisting of several steps; all are required for optimal results. The steps, shown in Figure 2-3, are as follows:

Planning—When the code compiles cleanly (no errors or warning messages), and after it passes through Lint (if used) the Author submits listings to the Moderator, who forms an inspection team. The Moderator distributes listings to each team member, as well as other related documents such as design requirements and documentation. The bulk of the Planning process is done by the Moderator, who can use email to coordinate with team members. An effective Moderator respects the time constraints of his or her colleagues and avoids interrupting them.

Overview—This *optional* step is a meeting when the inspection team members are not familiar with the development project. The Author pro-

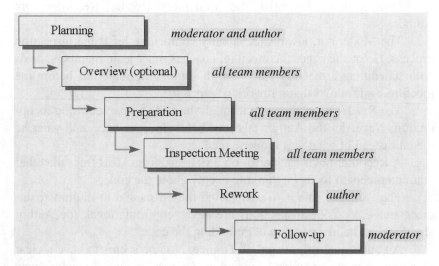

FIGURE 2-3 The Code Inspection process.

vides enough background to team members to facilitate their understanding of the code.

Preparation—Inspectors individually examine the code and related materials. They use a checklist to ensure that they check all potential problem areas. Each inspector marks up his or her copy of the code listing with suspected problem areas.

Inspection Meeting—The entire team meets to review the code. The Moderator runs the meeting tightly. The only subject for discussion is the code under review; any other subject is simply not appropriate and is not allowed.

The person designated as Reader presents the code by paraphrasing the meaning of small sections of code in a context higher than that of the code itself. In other words, the Reader is translating short code snippets from computer-lingo to English to ensure that the code's implementation has the correct meaning.

The Reader continuously decides how many lines of code to paraphrase, picking a number that allows reasonable extraction of meaning. Typically he's paraphrasing two or three lines at a time. He paraphrases every decision point, every branch, case, etc. One study concluded that only 50% of the code gets executed during typical tests, so be sure the inspection looks at *everything*.

Use a checklist to be sure you're looking at all important items. See the "Code Inspection Checklist" for details. Avoid ad hoc nitpicking; follow the firmware standard to guide all stylistic issues. Reject code that does not conform to the letter of the standard.

Log and classify defects as Major or Minor. A Major bug is one that could result in a problem visible to the customer. Minor bugs are those that include spelling errors, noncompliance with the firmware standards, and poor workmanship that does not lead to a major error.

Why the classification? Because when the pressure is on, when the deadline looms near, management will demand that you drop inspections as they don't seem like "real work." A list of classified bugs gives you the ammunition needed to make it clear that dropping inspections will yield more errors and slower delivery.

Fill out two forms. The "Code Inspection Checklist" is a summary of the number of errors of each type that are found. Use this data to understand the inspection process's effectiveness. The "Inspection Error List" contains the details of each defect requiring rework.

The code itself is the only thing under review; the author may not be criticized. One way to defuse the tension in starting up new inspection processes (before the team members are truly comfortable with it) is to have the Author supply a pizza for the meeting. Then he seems like the good guy.

At this meeting, make no attempt to rework the code or to come up with alternative approaches. Just find errors and log them; let the Author deal with implementing solutions. The Moderator must keep the meeting fast-paced and efficient.

Note that comment lines require as much review as code lines. Misspellings, lousy grammar, and poor communication of ideas are as deadly in comments as outright bugs in code. Firmware must *work*, and it must also *communicate its meaning*. The comments are a critical part of this and deserve as much attention as the code itself.

It's worthwhile to compare the size of the code to the estimate originally produced (if any!) when the project was scheduled. If it varies significantly from the estimate, figure out why, so you can learn from your estimation process.

Limit inspection meetings to a maximum of two hours. At the conclusion of the review of each function decide whether the code should be accepted as is or sent back for rework.

Rework—The Author makes all suggested corrections, gets a clean compile (and Lint if used) and sends it back to the Moderator.

Follow-up—The Moderator checks the reworked code. Once the Moderator is satisfied, the inspection is formally complete and the code may be tested.

Other Points

One hidden benefit of Code Inspections is their intrinsic advertising value. We talk about software reuse, while all too often failing spectacularly at it. Reuse is certainly tough, requiring lots of discipline. One reason reuse fails, though, is simply because people don't know a particular chunk of code exists. If you don't know there's a function on the shelf, ready to rock 'n' roll, then there's no chance you'll reuse it. When four people inspect code, four people have some level of buy-in to that software, and all four will generally realize the function exists.

The literature is full of the pros and cons of inspecting code before you get a clean compile. My feeling is that the compiler is nothing more than a tool, one that very cheaply and quickly picks up the stupid, silly errors we all make. Compile first and use a Lint tool to find other problems. Let the tools—not expensive people—pick up the simple mistakes.

I also believe that the only good compile is a clean compile. No error messages. No warning messages. Warnings are deadly when some other programmer, maybe years from now, tries to change a line. When presented with a screen full of warnings, he'll have no idea if these are normal or a symptom of a newly induced problem.

Do the inspection post-compile but pre-test. Developers constantly ask if they can do "a bit" of testing before the inspection—surely only to reduce the embarrassment of finding dumb mistakes in front of their peers. Sorry, but testing first negates most of the benefits. First, inspection is the cheapest way to find bugs; the entire point of it is to avoid testing. Second, all too often a pre-tested module never gets inspected. "Well, that sucker works OK; why waste time inspecting it?"

Tune your inspection checklist. As you learn about the types of defects you're finding, add those to the checklist so the inspection process benefits from actual experience.

Inspections work best when done quickly—but not too fast. Figure 2-4 graphs percentage of bugs found in the inspection versus number of lines inspected per hour as found in a number of studies. It's clear that at 500 lines per hour no bugs are found. At 50 lines per hour you're working inefficiently. There's a sweet spot around 150 lines per hour that detects most of the bugs you're going to find, yet keeps the meeting moving swiftly.

Code Inspections cannot succeed without a defined firmware standard. The two go hand in hand.

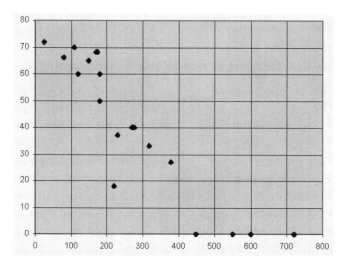

FIGURE 2-4 Percentage of bugs found versus number of lines inspected per hour.

> What does it cost to inspect code? We do inspections because they have a significant net *negative* cost. Yet sometimes management is not so sanguine; it helps to show the total cost of an inspection assuming there's *no* savings from downstream debugging.
>
> The inspection includes four people: the Moderator, Reader, Recorder, and Author. Assume (for the sake of discussion) that these folks average a $60,000 salary, and overhead at your company is 100%. Then:
>
> | | |
> |---|---|
> | One person costs: | $120,000 = $60,000 × 2 (overhead) |
> | One person costs: | $58/hr = $120,000/2080 work hours/year |
> | Four people cost: | $232/hr = $58/hr × 4 |
> | Inspection cost/line: | $1.54 = $232 per hour/150 lines inspected per hour |
>
> Since we know code costs $20–50 per line to produce, this $1.54 cost is obviously in the noise.

For more information on inspections, check out *Software Inspection*, Tom Gilb and Dorothy Graham, 1993, TJ Press (London), ISBN 0-201-63181-4, and *Software Inspection—An Industry Best Practice*, David Wheeler, Bill Brykczynski, and Reginald Meeson, 1996 by IEEE Computer Society Press (CA), ISBN 0-8186-7340-0.

Step 4: Create a Quiet Work Environment

For my money the most important work on software productivity in the last 20 years is DeMarco and Lister's *Peopleware* (1987, Dorset House Publishing, New York). Read this slender volume, then read it again, and then get your boss to read it.

For a decade the authors conducted coding wars at a number of different companies, pitting teams against each other on a standard set of software problems. The results showed that, using any measure of performance (speed, defects, etc.), the average of those in the first quartile outperformed the average in the fourth quartile by a factor of 2.6. Surprisingly, none of the factors you'd expect to matter correlated to the best and worst performers. Even experience mattered little, as long as the programmers had been working for at least 6 months.

Table 2-1 Code Inspection Checklist

Project:
Author:
Function Name:
Date:

Number of errors		Error type
Major	Minor	
		Code does not meet firmware standards
		Function size and complexity unreasonable
		Unclear expression of ideas in the code
		Poor encapsulation
		Function prototypes not correctly used
		Data types do not match
		Uninitialized variables at start of function
		Uninitialized variables going into loops
		Poor logic—won't function as needed
		Poor commenting
		Error condition not caught (e.g., return codes from malloc())?
		Switch statement without a default case (if only a subset of the possible conditions used)?
		Incorrect syntax—such as proper use of ==, =, &&, &, etc.
		Non-reentrant code in dangerous places
		Slow code in an area where speed is important
		Other
		Other

A Major bug is one that if not removed could result in a problem that the customer will see. Minor bugs are those that include spelling errors, non-compliance with the firmware standards, and poor workmanship that does not lead to a major error.

Table 2-2 Inspection Error List

Project:
Author:
Function Name:
Date:
Rework Required?

Location	Error description	Major	Minor

They did find a very strong correlation between the office environment and team performance. Needless interruptions yielded poor performance. The best teams had private (read "quiet") offices and phones with "off" switches. Their study suggests that quiet time saves vast amounts of money.

Think about this. The almost minor tweak of getting some quiet time can, according to their data, multiply your productivity by 260%! That's an astonishing result. For the same salary your boss pays you now, he'd get almost three of you.

The winners—those performing almost three times as well as the losers, had the following environmental factors:

	1st quartile	*4th quartile*
Dedicated workspace	78 sq ft	46 sq ft
Is it quiet?	57% yes	29% yes
Is it private?	62% yes	19% yes
Can you turn off phone?	52% yes	10% yes
Can you divert your calls?	76% yes	19% yes
Frequent interruptions?	38% yes	76% yes

Too many of us work in a sea of cubicles, despite the clear data showing how ineffective they are. It's bad enough that there's no door and no privacy. Worse is when we're subjected to the phone calls of all of our neighbors. We hear the whispered agony as the poor sod in the cube next door wrestles with divorce. We try to focus on our work . . . but because we're human, the pathos of the drama grabs our attention till we're straining to hear the latest development. Is this an efficient use of an expensive person's time?

> One correspondent told of working for a Fortune 500 company when heavy hiring led to a shortage of cubicles for incoming programmers. One was assigned a manager's office, complete with window. Everyone congratulated him on his luck. Shortly a maintenance worker appeared—and boarded up the window. The office police considered a window to be a luxury reserved for management, not engineers.
> Dysfunctional? You bet.

Various studies show that after an interruption it takes, on average, around 15 minutes to resume a "state of flow"—where you're once again deeply immersed in the problem at hand. Thus, if you are interrupted by colleagues or the phone three or four times an hour, *you cannot get any creative work done!* This implies that it's impossible to do support and development concurrently.

Yet the cube police will rarely listen to data and reason. They've invested in the cubes, and they've made a *decision*, by God! The cubicles are here to stay!

This is a case where we can only wage a defensive action. Try to educate your boss, but resign yourself to failure. In the meantime, take some action to minimize the downside of the environment. Here are a few ideas:

- Wear headphones and listen to music to drown out the divorce saga next door.
- Turn the phone off! If it has no "off" switch, unplug the damn thing. In desperate situations, attack the wire with a pair of wire cutters. Remember that a phone is a bell that anyone in the world can ring to bring you running. Conquer this madness for your most productive hours.
- Know your most productive hours. I work best before lunch; that's when I schedule all of my creative work, all of the hard stuff. I leave the afternoons free for low-IQ activities such as meetings, phone calls, and paperwork.
- Disable the email. It's worse than the phone. Your two hundred closest friends who send the joke of the day are surely a delight, but if you respond to the email reader's "bing" you're little more than one of NASA's monkeys pressing a button to get a banana.
- Put a curtain across the opening to simulate a poor man's door. Since the height of most cubes is rather low, use a Velcro fastener or a clip to secure the curtain across the opening. Be sure others understand that when it's closed you are not willing to hear from anyone unless it's an emergency.

An old farmer and a young farmer are standing at the fence talking about farm lore, and the old farmer's phone starts to ring. The old farmer just keeps talking about herbicides and hybrids, until the young farmer interrupts "Aren't you going to answer that?"

"What fer?" says the old farmer.

"Why, 'cause it's ringing. Aren't you going to get it?" says the younger.

The older farmer sighs and knowingly shakes his head. "Nope," he says. Then he looks the younger in the eye to make sure he understands, "Ya see, I bought that phone for *my* convenience."

Never forget that the phone is a bell that anyone in the world can ring to make you jump. Take charge of your time!

It stands to reason that we need to focus to think, and that we need to think to create decent embedded products. Find a way to get some privacy, and protect that privacy above all.

When I use the Peopleware argument with managers, they always complain that private offices cost too much. Let's look at the numbers.

DeMarco and Lister found that the best performers had an average of 78 square feet of private office space. Let's be generous and use 100. In the Washington, DC, area in 1998, nice—very nice—full-service office space runs around $30/square foot per year.

Cost: 100 square feet:	$3000/yr = 100 sq ft × $30/ft/year
One engineer costs:	$120,000 = $60,000 × 2 (overhead)
The office represents:	2.5% of cost of the worker = $3000/$120,000

Thus, if the cost of the cubicle is *zero*, then only a 2.5% increase in productivity pays for the office! Yet DeMarco and Lister claim a 260% improvement. Disagree with their numbers? *Even if they are off by an order of magnitude, a private office is 10 times cheaper than a cubicle.*

You don't have to be a rocket scientist to see the true cost/benefit of private offices versus cubicles.

Step 5: Measure Your Bug Rates

Code Inspections are an important step in bug reduction. But bugs—some bugs—will still be there. We'll never entirely eliminate them from firmware engineering.

Understand, though, that bugs are a natural part of software development. He who makes no mistakes surely writes no code. Bugs—or defects, in the parlance of the software engineering community—are to be expected. It's OK to make mistakes, as long as we're prepared to catch and correct these errors.

Though I'm not big on measuring things, bugs are such a source of trouble in embedded systems that we simply have to log data about them. There are three big reasons for bug measurements:

1. We find and fix them too quickly. We need to slow down and think more before implementing a fix. Logging the bug slows us down a trifle.
2. A small percentage of the code will be junk. Measuring bugs helps us identify these functions so we can take appropriate action.

3. Defects are a sure measure of customer-perceived quality. Once a product ships, we've got to log defects to understand how well our firmware processes satisfy the customer—the ultimate measure of success.

But first, a few words about "measurements."

It's easy to take data. With computer assistance we can measure just about anything and attempt to correlate that data to forces as random as the wind.

W. Edwards Deming, 1900–1993, quality-control expert, noted that using measurements as motivators is doomed to failure. He realized that there are two general classes of motivating factors: The first he called "intrinsic." These are things like professionalism, feeling like part of a team, and wanting to do a good job. "Extrinsic" motivators are those applied to a person or team, such as arbitrary measurements, capricious decisions, and threats. Extrinsic motivators drive out intrinsic factors, turning workers into uncaring automatons. This may or may not work in a factory environment, but is deadly for knowledge workers.

So measurements are an ineffective tool for motivation.

Good measures promote *understanding*. They transcend the details and reveal hidden but profound truths. These are the sorts of measures we should pursue relentlessly.

But we're all very busy and must be wary of getting diverted by the measurement process. Successful measures have the following three characteristics:

- They're easy to do.
- Each gives insight into the product and/or processes.
- The measure supports effective change-making. If we take data and do nothing with it, we're wasting our time.

For every measure, think in terms of first *collecting* the data, then *interpreting* it to make sense of the raw numbers. Then figure on *presenting* the data to yourself, your boss, or your colleagues. Finally, be prepared to *act* on the new understanding.

Stop, Look, Listen

In the bad old days of mainframes, computers were enshrined in technical tabernacles, serviced by a priesthood of specially vetted operators. Average users never saw much beyond the punch-card readers.

In those days of yore an edit–execute cycle started with punching perhaps thousands of cards, hauling them to the computer center (being careful not to drop the card boxes; on more than one occasion I saw grad students break down and weep as they tried to figure out how to order the cards splashed across the floor), and then waiting a day or more to see how the run went. Obviously, with a cycle this long, no one could afford to use the machine to catch stupid mistakes. We learned to "play computer" (sadly, a lost art) to deeply examine the code before the machine ever had a go at it.

How things have changed! Found a bug in your code? No sweat—a quick edit, compile, and re-download takes no more than a few seconds. Developers now look like hummingbirds doing a frenzied edit–compile–download dance.

It's wonderful that advancing technology has freed us from the dreary days of waiting for our jobs to run. Watching developers work, though, I see we've created an insidious invitation to bypass *thinking*.

How often have you found a problem in the code, and thought, "Uh, if I change this, maybe the bug will go away?" To me that's a sure sign of disaster. If the change fails to fix the problem, you're in good shape. The peril is when a poorly thought-out modification does indeed "cure" the defect. Is it really cured? Or did you just mask it?

Unless you've thought things through, *any* change to the code is an invitation to disaster.

Our fabulous tools enable this dysfunctional pattern of behavior. To break the cycle we have to slow down a bit.

EEs traditionally keep engineering notebooks, bound volumes of numbered pages, ostensibly for patent protection reasons but more often useful for logging notes, ideas, and fixes. Firmware folks should do no less.

When you run into a problem, stop for a few seconds. Write it down. Examine your options and list those as well. Log your proposed solution (see Figure 2-5).

Keeping such a journal helps force us to think things through more clearly. It's also a chance to reflect for a moment, and, if possible, come up with a way to avoid that sort of problem in the future.

> One colleague recently fought a tough problem with a wild pointer. While logging the symptoms and ideas for fixing the code, he realized that this particular flavor of bug could appear in all sorts of places in the code. Instead of just plodding on, he set up a logic analyzer to trigger on the wild writes . . . and found seven other areas with the same problem, all of which had not as yet exhibited a symptom. Now that's what I call a great debug strategy—using experience to predict problems!

> **C-MON PROJECT**
> 3-10-97
>
> BUG: STATUS LED ALWAYS OFF
> PROBLEM: STATUS TASK NEVER DISPATCHED!
> FIX: ADD TASK TO RTOS INITIALIZE
>
> BUG: STUPID STATUS LED REVERSED SENSE!!
> PROBLEM: JLE SHOULD BE JGE INSTRUCTION. I ALWAYS GET THAT WRONG.
> FIX: CHANGE INSTRUCTION AND ALSO ADD COMMENT TO PREVIOUS COMPARE SO TO MAKE OPERATION MORE OBVIOUS: DESTINATION – SOURCE (NOT REVERSE!)

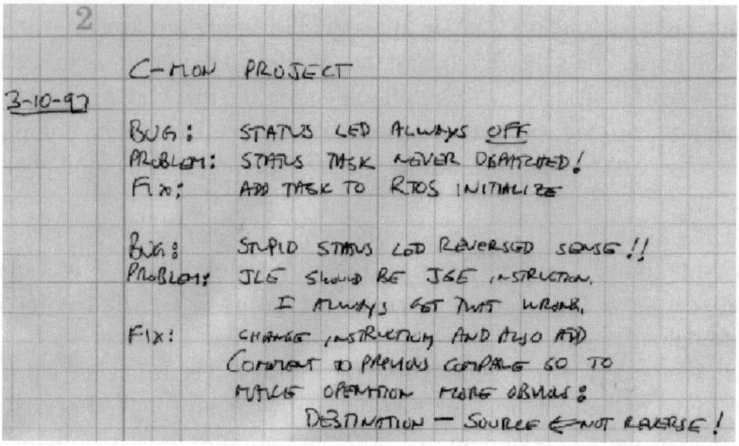

FIGURE 2-5 A personal bug log.

Identify Bad Code

Barry Boehm found that typically 80% of the defects in a program are in 20% of the modules. IBM's numbers showed that 57% of the bugs are in 7% of modules. Weinberg's numbers are even more compelling: 80% of the defects are in 2% of the modules.

In other words, *most of the bugs will be in a few modules or functions*. These academic studies confirm our common sense. How many times have you tried to beat a function into submission, fixing bug after bug after bug, convinced that this one is (you hope!) the last?

We've all also had that awful function that just simply stinks. It's ugly. The one that makes you slightly nauseous every time you open it. A decent Code Inspection will detect most of these poorly crafted beasts, but if one slips through, we have to take some action.

Make identifying bad code a priority. Then trash those modules and start over.

It sure would be nice to have the chance to write every program twice: the first time to gain a deep understanding of the problem; the second to do it right. Reality's ugly hand means that's not an option. But the bad code, the code where we spend far too much time debugging, needs to be excised and redone. The data suggests we're talking about recoding only around 5% of the functions—not a bad price to pay in the pursuit of quality.

Boehm's studies show that these problem modules cost, on average, *four times* as much as any other module. So, if we identify these modules (by tracking bug rates), we can rewrite them *twice* and still come out ahead!

Step 6: Measure Your Code Production Rates

Schedules collapse for a lot of reasons. In the 50 years people have been programming electronic computers, we've learned one fact above all: without a clear project specification, any schedule estimate is nothing more than a stab in the dark. Yet every day dozens of projects start with little more definition than, "Well, build a new instrument kind of like the last one, with more features, cheaper, and smaller." Any estimate made to a vague spec is totally without value.

The corollary is that given the clear spec, we need time—sometimes *lots* of time—to develop an accurate schedule. It ain't easy to translate a spec into a design, and then to realistically size the project. You simply cannot do justice to an estimate in two days, yet that's often all we get.

Further, managers must accept schedule estimates made by their people. Sure, there's plenty of room for negotiation: reduce features, add resources, or permit more bugs (gasp!). Yet most developers tell me their schedule estimates are capriciously changed by management to reflect a desired end date, with no corresponding adjustments made to the project's scope.

The result is almost comical to watch, in a perverse way. Developers drown themselves in project management software, mousing milestone triangles back and forth to meet an arbitrary date cast in stone by management. The final printout may look encouraging, but generally gets the total lack of respect it deserves from the people doing the actual work. The schedule is then nothing more than dishonesty codified as policy.

There's an insidious sort of dishonest estimation too many of us engage in. It's easy to blame the boss for schedule debacles, yet often we bear plenty of responsibility. We get lazy, and we don't invest the same amount of thought, time, and energy into scheduling that we give to debugging. "Yeah, that section's kind of like something I did once before" is, at best, just a start of estimation. You cannot derive time, cost, or size from such a vague statement . . . yet too many of us do. "Gee, that looks pretty easy—say a week" is a variant on this theme.

Doing less than a thoughtful, thorough job of estimation is a form of self-deceit that rapidly turns into an institutionalized lie. "We'll ship December 1," we chant, while the estimators know just how flimsy the framework of that belief is. Marketing prepares glossy brochures, technical pubs writes the manual, and production orders parts. December 1 rolls around, and, surprise! January, February, and March go by in a blur. Eventually the product goes out the door, leaving everyone exhausted and angry. Too much of this stems from a lousy job done in the first week of the project when we didn't carefully estimate its complexity.

It's time to stop the madness!

We learn in school to practice top-down decomposition. Design the system, break each block into smaller chunks, and iterate until no part of the code is more than a page or two long. Then, and only then, can you understand its complexity. We generally then take a reasonable guess: "This module will be 50 lines of code." (Instead of lines of code, some companies use function points or other units of measure.)

Swell. Do this and you will still almost certainly fail.

Few developers seem to understand that knowing code size—even if it were 100% accurate—is only half of the data absolutely required to produce any kind of schedule. It's amazing that somehow we manage to solve the equation

$$\text{development time} = (\text{program size in Lines of Code}) \times (\text{time per Line of Code})$$

when time-per-Line-of-Code is totally unknown.

If you estimate modules in terms of lines of code (LOC), then you must know—exactly—the cost per LOC. Ditto for function points or any other unit of measure. Guesses are not useful.

When I sing this song to developers, the response is always, "Yeah, sure, but I don't have LOC data . . what do I do about the project I'm on today?" There's only one answer: sorry, pal—you're outta luck. IBM's LOC/month number is useless to you, as is one from the FAA, DOD, or any other organization. In the commercial world we all hold our code to different standards, which greatly skews productivity in any particular measure.

You simply must measure how fast you generate embedded code, every single day, for the rest of your life. It's like being on a diet—even when everything's perfect, and you've shed those 20 extra pounds, you'll forever be monitoring your weight to stay in the desired range. Start collecting the data today, do it forever, and over time you'll find a model of your productivity that will greatly improve your estimation accuracy. Don't do it, and every estimate you make will be, in effect, a lie—a wild, meaningless guess.

Step 7: Constantly Study Software Engineering

The last step is the most important. Study constantly. In the 50 years since ENIAC we've learned a lot about the right and wrong ways to build software; almost all of the lessons are directly applicable to firmware development.

How does an elderly, near-retirement doctor practice medicine? In the same way he did before World War II, before penicillin? Hardly. Doctors spend a lifetime learning. They understand that lunch time is always spent with a stack of journals.

Like doctors, we practice in a dynamic, changing environment. Unless we master better ways of producing code we'll be the metaphorical equivalent of the sixteenth-century medicine man, trepanning instead of practicing modern brain surgery.

Learn new techniques. Experiment with them. Any idiot can write code; the geniuses are those who find better ways of writing code.

> One of the more intriguing approaches to creating a discipline of software engineering is the Personal Software Process, a method created by Watts Humphrey. An original architect of the CMM, Humphrey realized that developers need a method they can use *now*, without waiting for the CMM revolution to take hold at their company. His vision is not easy, but the benefits are profound. Check out his *A Discipline for Software Engineering,* Watts S. Humphrey, 1995, Addison-Wesley.

Summary

With a bit of age (but less than anticipated maturity), it's interesting to look back and to see how most of us form personalities very early in life, personalities with strengths and weaknesses that largely stay intact over the course of decades.

The embedded community is composed of mostly smart, well-educated people, many of whom believe in some sort of personal improvement. But, are we successful? How many of us live up to our New Year's resolutions?

Browse any bookstore. The shelves groan under self-help books. How many people actually get helped, or at least helped to the point of being done with a particular problem? Go to the diet section—I think there are more diets being sold than the sum total of national excess pounds. People buy these books with the best of intentions, yet every year America gets a little heavier.

Our desires and plans for self-improvement—at home or at the office—are among the more noble human characteristics. The reality is that we fail—a lot. It seems the most common way to compensate is a promise made to ourselves to "try harder" or to "do better." It's rarely effective.

Change works best when we change the way we do things. Forget the vague promises—invent a new way of accomplishing your goal. Planning on reducing your drinking? Getting regular exercise? Develop a process that ensures that you're meeting your goal.

The same goes for improving your abilities as a developer. Forget the vague promises to "read more books" or whatever. Invent a solution that has a better chance of succeeding. Even better—steal a solution that works from someone else.

Cynicism abounds in this field. We're all self-professed experts of development, despite the obvious evidence of too many failed projects.

I talk to a lot of companies who are convinced that change is impossible; that the methods I espouse are not effective (despite the data that shows the contrary), or that "management" will never let them take the steps needed to effect change.

That's the idea behind the "7 Steps." Do it covertly, if need be; keep management in the dark if you're convinced of their unwillingness to use a defined software process to create better embedded projects faster.

If management is enlightened enough to understand that the firmware crisis requires change—and lots of it!—then educate them as you educate yourself.

Perhaps an analogy is in order. The industrial revolution was spawned by a lot of forces, but one of the most important was the concentration of capital. The industrialists spent vast sums on foundries, steel mills, and other means of production. Though it was possible to hand-craft cars, dumping megabucks into assembly lines and equipment yielded lower prices, and eventually paid off the investment in spades.

The same holds true for intellectual capital. Invest in the systems and processes that will create massive dividends over time. If we're unwilling to do so, we'll be left behind while others, more adaptable, put a few bucks up front and win the software wars.

> A final thought:
> If you're a process cynic, if you disbelieve all I've said in this chapter, ask yourself one question: do I *consistently* deliver products on time and on budget?
> If the answer is no, then *what are you doing about it?*

CHAPTER **3**

Stop Writing Big Programs

The most important rule of software engineering is also the least known: *Complexity does not scale linearly with size.*

For "complexity" substitute any difficult parameter, such as time required to implement the project, bugs, or how well the final product meets design specifications (unhappily, meeting design specs is all too often uncorrelated with meeting customer requirements . . .).

So a 2000-line program requires *more* than twice as much development time as one that's half the size.

A bit of thought confirms this. Surely, any competent programmer can write an utterly perfect five-line program in 10 minutes. Multiply the five lines and the 10 minutes by a hundred; those of us with an honest assessment of our own skills will have to admit the chances of writing a perfect 500 line program in 16 hours are slim at best.

Data collected on hundreds of IBM projects confirm this. As systems become more complex they take longer to produce, both because of the extra size and *because productivity falls dramatically:*

(man-yrs)	Lines of code produced per month
1	439
10	220
100	110
1000	55

Look closely at this data. Notice that there's an order of magnitude increase in delivery time simply due to the reduced productivity as the project's magnitude swells.

35

COCOMO Data

Barry Boehm codified this concept in his Constructive Cost Model (COCOMO). He found that

$$\text{Effort to create a project} = C \times \text{KLOC}^M.$$

(KLOC means "thousands of lines of code.")

Though the exact values of C and M vary depending on a number of factors (e.g., real-time code is harder than that for the user interface), both are always greater than 1.

A bit of algebra shows that, since $M > 1$, effort grows much faster than the size of the program.

For real-time projects managed with the very best practices, C is typically 3.6 and M around 1.2. In embedded systems, which combine the worst problems of real time with hardware dependencies, these coefficients are higher. Toss in the typical poor software practices of the embedded industries and the M exponent can climb well above 1.5.

Suppose $C = 1$ and $M = 1.4$. At the risk of oversimplifying Boehm's model, we can still get an idea of the nonlinear growth of complexity with program size as follows:

Lines of code	Effort	Comments
10,000	25.1	
20,000	66.3	Double size of code; effort goes up by 2.64
100,000	631	Size grows by factor of 10; effort grows by 25

So, in doubling the size of the program we incur 32% *additional* overhead.

The human analogy of this phenomenon is the one so colorfully illustrated by Fred Brooks in his *The Mythical Man-Month* (a must read for all software folks). As projects grow, adding people has a diminishing return. One reason is the increased number of communications channels. Two people can only talk to each other; there's only a single comm path. Three workers have three communications paths; four have six. In fact, the growth of links is exponential: given n workers, there are $(n^2 - n)/2$ links between team members.

In other words, add one worker and suddenly he's interfacing in n^2 ways with the others. Pretty soon memos and meetings eat up the entire work day.

The solution is clear: break teams into smaller, autonomous, and independent units to reduce these communications links.

Similarly, cut programs into smaller units. Since a large part of the problem stems from dependencies (global variables, data passed between functions, shared hardware, etc.), find a way to partition the program to eliminate—or minimize—the dependencies between units.

Traditional computer science would have us believe the solution is top-down decomposition of the problem, perhaps then encapsulating each element into an OOP object. In fact, "top-down design," "structured programming," and "OOP" are the holy words of the computer vocabulary; like fairy dust, if we sprinkle enough of this magic on our software all of the problems will disappear.

I think this model is one of the most outrageous scams ever perpetrated on the embedded community. Top-down design and OOP are wonderful concepts, but are nothing more than a subset of our arsenal of tools.

I remember interviewing a new college graduate, a CS major. It was eerie, really, rather like dealing with a programmed cult member unthinkingly chanting the persuasion's mantra. In this case, though, it was the tenets of structured programming mindlessly flowing from his lips.

It struck me that programming has evolved from a chaotic "make it work no matter what" level of anarchy to a pseudo-science whose precepts are practiced without question. Problem Analysis, Top-Down Decomposition, OOP—all of these and more are the commandments of structured design, commandments we're instructed to follow lest we suffer the pain of failure.

Surely there's room for iconoclastic ideas. I fear we've accepted structured design, and all it implies, as a bedrock of our civilization, one buried so deep we never dare to wonder if it's only a part of the solution.

Top-down decomposition and OOP design are merely screwdrivers or hammers in the toolbox of *partitioning* concepts.

Partitioning

Our goal in firmware design is to cheat the exponential in the CO-COMO model, the exponential that also shows up in every empirical study of software productivity. We need to use every conceivable technique to flatten the curve, to move the M factor close to unity.

Top-down decomposition is a useful weapon in cheating the COCOMO exponential, as is OOP design. In embedded systems we have other possibilities denied to many people building desktop applications.

Partition with Encapsulation

The OOP advocates correctly and profoundly point out the benefit of encapsulation, to my mind the most important of the tripartite mantra *encapsulation, inheritance, and polymorphism.*

Above all, encapsulation means binding functions together with the functions' data. It means hiding the data so no other part of the program can monkey with it. All access to the data takes place through function calls, not through global variables.

Instead of reading a status word, your code calls a status function. Rather than diddle a hardware port, you insulate the hardware from the code with a driver.

Encapsulation works equally well in assembly language or in C++ (Figure 3-1). It requires a *will to bind data with functions* rather than any particular language feature. C++ will not save the firmware world; encapsulation, though, is surely part of the solution.

One of the greatest evils in the universe, an evil in part responsible for global warming, ozone depletion, and male pattern baldness, is the use of global variables.

What's wrong with globals? A partial list includes:

- Any function, anywhere in the program, can change a global variable at will. This makes finding why a global change is a nightmare. Without the very best of tools you'll spend too much time finding simple bugs; time invested chasing problems will be all out of proportion to value received.
- Globals create tremendous reentrancy problems, as we'll see in Chapter 4.
- While distance may make the heart grow fonder, it also clouds our memories. A huge source of bugs is assigning data to variables defined in a remote module with the wrong type, or over- and under-running buffers as we lose track of their size, or forgetting to null-terminate strings. If a variable is defined in its referring code, it's awfully hard to forget type and size info.

Every firmware standard—backed up by the rigorous checks of code inspections—must set rules about global use. Though we'd like to ban them entirely, the truth is that in real-time systems they are sometimes unavoidable. Nothing is faster than a global flag; when speed is truly an issue, a few, a very few, globals may indeed be required. Restrict their use to only a few critical areas. I feel that defining a global is such a source of problems that the team leader should approve every one.

```
_text         segment
;
; _get_cba_min—read a min value at (index) from the
; CBA buffer. Called by a C program with the (index)
; argument on the stack.
;
; Returns result in AX.
;
  public _get_cba_min
_get_cba_min proc far
    mov     bx,sp
    mov     bx,[bx+4]       ; bx= index in buf to read
    add     bx,cba_buf      ; add offset to make addr
    push    ds
    mov     dx,buffer_seg   ; point to the buffer seg
    mov     es,dx
    mov     ax,es:bx        ; read the min value
    pop     ds
    retf
    endp
_text         ends
;
; CBA buffer, which is managed by the *_cba routines.
; Format:   100 entries, each of which looks like:
;     buf+0   min value (word)
;     buf+2   max value (word)
;     buf+4   number of iterations (word)
;
_data         segment para 'DATA'
cba_buf       ds              100 * 6     ; CBA buffer
_data         ends
```

FIGURE 3-1 Encapsulation in assembly language. Note that the data is not defined Public.

Among the great money-makers for ICE vendors are complex hardware breakpoints, used most often for chasing down errant changes to global variables. If you like globals, figure on anteing up plenty for tools.

There's yet one more waffle on my anti-global crusade: device handlers sometimes must share data stored in common buffers and the like. We do not write a serial receive routine in isolation. It's part of a fabric of handlers that include input, output, initialization, and one or more interrupt service routines (ISRs).

This implies something profound about module design. Write programs with lots and lots of modules! Don't lump code into a handful of 5000-line files. Assign one module per logical function: for example, have a single module (file) that includes all of the serial device handlers—*and nothing else*. Structurally it looks like:

```
                public    serial_in, serial_out,
serial_init
serial_in:        code
serial_out:       code
serial_init:      code
serial_isr:       code
                private   data
buffer:    data
status:    data
```

The data items are filescopics—global to the module but private to the rest of the system. I feel this tradeoff is needed in embedded systems to reduce performance penalties of the noble but not-always-possible anti-global tack.

Partition with CPUs

Given that firmware is the most expensive thing in the universe, given that the code will always be the most expensive part of the development effort, given that we're under fire to deliver more complex systems to market faster than ever, it makes sense in all but the most cost-sensitive systems to have the hardware design fall out of software considerations. That is, design the hardware in a way to minimize the cost of software development.

It's time to reverse the conventional design approach, and *let the software drive the hardware design.*

Consider the typical modern embedded system. A single CPU has the metaphorical role of a mainframe computer: it handles all of the inputs and outputs, runs application code, and services interrupts. Like the mainframe, one CPU, one program, is doing many disparate activities that only eventually serve a common goal.

Not enough horsepower? Toss in a 32-bitter. Crank up the clock rate. Cut out wait states.

Why do we continue to emulate the antiquated notion of "big iron"— even if the central machine is only an 8051? Mainframes were long ago replaced by distributed workstations.

A single big CPU running the entire application implies that there's a huge program handling everything. We know that big programs are bad—they cost too much to develop.

It's usually cheaper to add more CPUs merely for the sake of simplifying the software.

In the following table, "Effort" refers to development time as predicted by the COCOMO metric. The first two columns show the effort required to produce a single-CPU chunk of firmware of the indicated number of lines of code. The next five columns show models of partitioning the code over multiple CPUs—a "main" processor that runs the bulk of the application code, and a number of quite small "extra" microcontrollers for handling peripherals and similar tasks.

Single CPU		Multiple CPUs						
LOC	Effort	Main LOC	LOC/extra CPU	# extra CPUs	Total LOC	Effort	Faster[1]	Faster[2]
10,000	25	6000	2500	2	11000	19	22%	37%
20,000	66	12000	2500	4	22000	47	29%	48%
50,000	239	24000	5000	6	54000	143	40%	57%
100,000	631	50000	5000	12	110000	353	44%	65%

Clearly, total effort to produce the system decreases quite rapidly when tasks are farmed out to additional processors, even though these numbers include about 10% extra overhead to deal with interprocessor communication. The "Faster[1]" column shows how much faster we can deliver the system as a result.

But the numbers are computed using an exponent of 1.4 for M, which is a result of creating a big, complicated real-time embedded system. It's reasonable to design a system with as few real-time constraints as possible in the main CPU, allocating these tasks to the smaller and more tractable extra controllers. If we then reduce M to 1.2 for the main CPU (Boehm's real-time number) and leave it at 1.4 for the smaller processors that are working with fickle hardware, the numbers in the Faster[2] column result.

To put this in another context, getting a 100K LOC program to market 65% faster means we've saved over 200 man-months of development (using the fastest of Bell Lab's production rates), or something like $2 million.

Don't believe me? Cut the numbers by a factor of 10. That's still $200,000 in engineering that does not have to get amortized into the cost of the product. The product also gets to market much, much faster, and ideally it generates substantially more sales revenue.

The goal is to flatten the curve of complexity. Figure 3-2 shows the relative growth rates of effort—normalized to program size—for both approaches.

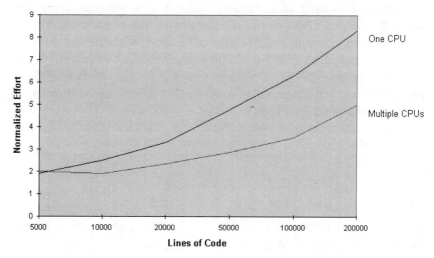

FIGURE 3-2 Flattening the curve of complexity growth.

NRE versus COGS

Nonrecurring engineering costs (NRE costs) are the bane of most technology managers' lives. NRE is that cost associated with developing a product. Its converse is the cost of goods sold (COGS), a.k.a. recurring costs.

NRE costs are amortized over the life of a product in fact or in reality. Mature companies carefully compute the amount of engineering in the product—a car maker, for instance, might spend a billion bucks engineering a new model with a lifespan of a million units sold; in this case the cost of the car goes up by $1000 to pay for the NRE. Smaller technology companies often act like cowboys and figure that NRE is just the cost of doing business; if we are profitable, then the product's price somehow (!) reflects all engineering expenses.

Increasing NRE costs drives up the product's price (most likely making it less competitive and thus reducing profits), or directly reduces profits.

Making an NRE versus COGS decision requires a delicate balancing act that deeply mirrors the nature of your company's product pricing. A $1 electronic greeting card cannot stand any extra components; minimize COGS above all. In an automobile the quantities are so large that engineers agonize over saving a foot of wire. The converse is a one-off or short-production-run device. The slightest development hiccup costs tens of thousands—easily—which will have to be amortized over a very small number of units.

Sometimes it's easy to figure the tradeoff between NRE and COGS. You should also consider the extra complication of opportunity costs—"If I do this, then what is the cost of not doing that?" As a young engineer I realized that we could save about $5000 a year by changing from EPROMS to masked ROMs. I prepared a careful analysis and presented it to my boss, who instantly turned it down because making the change would shut down my other engineering activities for some time. In this case we had a tremendous backlog of projects, any of which could yield more revenue than the measly $5K saved. In effect, my boss's message was, "You are more valuable than what we pay you." (That's what drives entrepreneurs into business—the hope they can get the extra money into their own pockets!)

Follow these guidelines to be successful in simplifying software through multiple CPUs:

- Break out nasty real-time hardware functions into independent CPUs. Do interrupts come at 1000/second from a device? Partition it to a controller and offload all of that ISR overhead from the main processor.
- Think microcontrollers, not microprocessors. Controllers are inherently limited in address space, which helps keep firmware size under control. Controllers are cheap (some cost less than 40 cents in quantity). Controllers have everything you need on one chip—RAM, ROM, I/O, etc.

- Think OTP—one-time programmable—or EEROM memory. Both let you build and test the application without going to expensive masked ROM. Quick to build, quick to burn, and quick to test.
- Keep the size of the code in the microcontrollers small. A few thousand lines is a nice, tractable size that even a single programmer working in isolation can create.
- Limit dependencies. One beautiful benefit of partitioning code into controllers is that you're pin-limited—the handful of pins on the chips acts as a natural barrier to complex communications and interaction between processors. Don't defeat this by layering a hideous communications scheme on top of an elegant design.

Communications is always a headache in multiple-processor applications. Building a reliable parallel comm scheme beats Freddy Krueger for a nightmare any day. Instead, use a standard, simple protocol such as I^{2}C. This is a two-wire serial protocol supported directly by many controllers. It's multi-master and multi-slave, so you can hang many processors on one pair of I^{2}C wires. With rates to 1 Mb/sec, there's enough speed for most applications. Even better: you can steal the code from Microchip's and National Semiconductor's Web sites.

The hardware designers will object to adding processors, of course. Just as firmware folks take pride in producing optimum code, our hardware brethren, too, want an elegant, minimalist creation where there's enough logic to make the thing work, but nothing more. Adding hardware—which has a cost—just to simplify the code seems like a terrible waste of resources.

Yet we've been designing systems with extra hardware for decades. There's no reason we couldn't build a software implementation of a UART. "Bit banging" software has been around for years. Instead, most of the time we'll add the UART device to eliminate the nasty, inefficient software solution.

One of Xerox's copiers is a monster of a machine that does everything but change the baby. An older design, it uses seven 8085s tied together with a simple proprietary network. One handles the paper mechanism, another the user interface, yet another error processing. The boards are all pretty much the same, and no ROM exceeds 32k. The machine is amazingly complex and feature-rich . . . but code sizes are tiny.

Partition by Features

Carpenters think in terms of studs and nails, hammers and saws. Their vision is limited to throwing up a wall or a roof. An architect, on the other hand, has a vision that encompasses the entire structure—but more importantly, one that includes a focus on the customer. The only meaningful measure of the architect's success is his customer's satisfaction.

We embedded folks too often distance ourselves from the customer's wants and needs. A focus on cranking schematics and code will thwart us from making the thousands of little decisions that transcend even the most detailed specification. *The only view of the product that is meaningful is the customer's.* Unless we think like the customer, we'll be unable to satisfy him. A hundred lines of beautiful C or 100k of assembly—it's all invisible to the people who matter most.

Instead of analyzing a problem entirely in terms of functions and modules, look at the product in the feature domain, since features are the customer's view of the widget. Manage the software using a matrix of features.

Table 3-1 shows the feature matrix for a printer. Notice that the first few items are not really features; they're basic, low-level functions required just to get the thing to start up, as indicated by the "Importance" factor of "required."

Beyond these, though, are things used to differentiate the product from competitive offerings. Downloadable fonts might be important, but do not affect the unit's ability to just put ink on paper. Image rotation, listed as the least important feature, sure is cool, but may not always be required.

Table 3-1

Feature	Importance	Priority	Complexity
Shell	Required		500
RTOS	Required		(purchased)
Keyboard handler	Required		300
LED driver	Required		500
Comm with host	Required		4,000
Paper handling	Required		2,000
Print engine	Required		10,000
Downloadable fonts	Important	1	1,000
Main 100 local fonts	Important	2	6,000
Unusual local fonts	Less important	3	10,000
Image rotation	Less important	4	3,000

The feature matrix ensures we're all working on the right part of the project. Build the important things first! Focus on the basic system structure—get all of it working, perfectly—before worrying about less important features. I see project after project in trouble because the due date looms with virtually nothing complete. Perhaps hundreds of functions work, but the unit cannot do anything a customer would find useful. Developers' efforts are scattered all over the project so that until everything is done, nothing is done.

The feature matrix is a scorecard. If we adopt the view that we're working on the important stuff first, and that until a feature works perfectly we do not move on, then any idiot—including those warming seats in marketing—can see and understand the project's status.

(The complexity rating shown is in estimated lines of code. LOC as a unit of measure is constantly assailed by the software community. Some push function points—unfortunately there are a dozen variants of this—as a better metric. Most often people who rail against LOC as a measure in fact measure nothing at all. I figure it's important to measure something, something easy to count, and LOC gives a useful if less than perfect assessment of complexity.)

Most projects are in jeopardy from the outset, as they're beset by a triad of conflicting demands (Figure 3-3). Meeting the schedule, with a high-quality product, that does everything the 24-year-old product manager in marketing wants, is usually next to impossible.

Eighty percent of all embedded systems are delivered late. Lots and lots of elements contribute to this, but we too often forget that when developing a product we're balancing the schedule/quality/features mix. Cut enough features and you can ship today. Set the quality bar to near zero

FIGURE 3-3 The twisted tradeoff.

and you can neglect the hard problems. Extend the schedule to infinity and the product can be perfect and complete.

Too many computer-based products are junk. Companies die or lose megabucks as a result of prematurely shipping something that just does not work. Consumers are frustrated by the constant need to reset their gadgets and by products that suffer the baffling maladies of the binary age.

We're also amused by the constant stream of announced-but-unavailable products. Firms do quite exquisite PR dances to explain away the latest delay; Microsoft's renaming of a late Windows upgrade to "95" bought them an extra year and the jeers of the world. Studies show that getting to market early reaps huge benefits; couple this with the extreme costs of engineering and it's clear that "ship the damn thing" is a cry we'll never cease to hear.

Long-term success will surely result from shipping a *quality* product *on time*. That means there's only one leg of the twisted tradeoff left to fiddle. Cut a few of the less important features to get a first-class device to market fast.

The computer age has brought the advent of the feature-rich product that no one understands or uses. My cell phone's "Function" key takes a two-digit argument—one hundred user-selectable functions/features built into this little marvel. Never use them, of course. I wish the silly thing could reliably establish a connection! The design team's vision was clearly skewed in term of features over quality, to consumers' loss.

If we're unwilling to partition the product by features, and to build the firmware in a clear, high-priority features-first hierarchy, we'll be forever trapped in an impossible balance that will yield either low quality or late shipment. Probably both.

Use a feature matrix, implementing each in a logical order, and *make each one perfect before you move on*. Then at any time management can make a reasonable decision: ship a quality product now, with this feature mix, or extend the schedule until more features are complete.

This means you must break down the code by feature, and only then apply top-down decomposition to the components of each feature. It means you'll manage by feature, getting each done before moving on, to keep the project's status crystal clear and shipping options always open.

Management may complain that this approach to development is, in a sense, planning for failure. They want it all: schedule, quality, and features. *This is an impossible dream!* Good software practices will certainly help hit all elements of the triad, but we've got to be prepared for problems.

Management uses the same strategy in making their projections. No wise CEO creates a cash flow plan that the company must hit to survive;

there's always a backup plan, a fall-back position in case something unexpected happens.

So, while partitioning by features will not reduce complexity, it leads to an earlier shipment with less panic as a workable portion of the product is complete at all times.

In fact, this approach suggests a development strategy that maximizes the visibility of the product's quality and schedule.

Develop Firmware Incrementally

Deming showed the world that it's impossible to test quality into a product. Software studies further demonstrate the futility of expecting test to uncover huge numbers of defects in reasonable times—in fact, some studies show that up to 50% of the code may never be exercised under a typical test regime.

Yet test is a necessary part of software development.

Firmware testing is dysfunctional and unlikely to be successful when postponed till the end of the project. The panic to ship overwhelms common sense; items at the end of the schedule are cut or glossed over. Test is usually a victim of the panic.

Another weak point of all too many schedules is that nasty line item known as "integration." Integration, too, gets deferred to the point where it's poorly done.

Yet integration shouldn't even exist as a line item. Integration implies we're only fiddling with bits and pieces of the application, ignoring the problem's gestalt, until very late in the schedule when an unexpected problem (unexpected only by people who don't realize that the reason for test is to unearth unexpected issues) will be a disaster.

The only reasonable way to build an embedded system is to start integrating today, now, on the day you first crank a line of code. The biggest schedule killers are unknowns; only testing and actually running code and hardware will reveal the existence of these unknowns.

As soon as practicable, build your system's skeleton and switch it on. Build the startup code. Get chip selects working. Create stub tasks or calling routines. Glue in purchased packages and prove to yourself that they work as advertised and as required. Deal with the vendor, if trouble surfaces, *now* rather than in a last-minute debug panic when they've unexpectedly gone on holiday for a week.

This is a good time to slip in a ROM monitor, perhaps enabled by a secret command set. It'll come in handy when you least have time to add one—perhaps in a panicked late-night debugging session moments before shipping, or for diagnosing problems that creep up in the field.

In a matter of days or a week or two you'll have a skeleton assembled, a skeleton that actually operates in some very limited manner. Perhaps it runs a null loop. Using your development tools, test this small scale chunk of the application.

Start adding the lowest-level code, testing as you go. Soon your system will have all of the device drivers in place (tested), ISRs (tested), the startup code (tested), and the major support items such as comm packages and the RTOS (again tested). Integration of your own applications code can then proceed in a reasonably orderly manner, plopping modules into a known-good code framework, facilitating testing at each step.

The point is to immediately build a framework that operates, and then drop features in one at a time, testing each as it becomes available. You're testing the entire system, such as it is, and expanding those tests as more of it comes together. Test and integration are no longer individual milestones; they are part of the very fabric of development.

Success requires a determination to constantly test. Every day, or at least every week, build the entire system (using all of the parts then available) and ensure that things work correctly. *Test constantly.* Fix bugs immediately.

The daily or weekly testing is the project's heartbeat. It ensures that the system really can be built and linked. It gives a constant view of the system's code quality, and encourages early feature feedback (a mixed blessing, admittedly—but our goal is to satisfy the customer, even at the cost of accepting slips due to reengineering poor feature implementation).

At the risk of sounding like a new-age romantic, someone working in aromatherapy rather than pushing bits around, we've got to learn to deal with human nature in the design process. Most managers would trade their firstborn for an army of Vulcan programmers, but until the Vulcan economy collapses ("emotionless programmer, will work for peanuts and logical discourse"), we'll have to find ways to efficiently use humans, with all of their limitations.

We people need a continuous feeling of accomplishment to feel effective and to be effective. Engineering is all about *making things work*; it's important to recognize this and create a development strategy that satisfies this need. Having lots of little progress points, where we see our system doing something, is tons more satisfying than coding for a year before hitting the ON switch.

A hundred thousand lines of carefully written and documented code is nothing more than worthless bits until it's tested. We hear "It's done" all the time in this field, where "done" might mean "vaguely understood" or "coded." To me "done" has one meaning only: "tested."

Incremental development and testing, especially of the high-risk areas such as hardware and communications, reduces risk tremendously. Even when we're not honest with each other ("Sure, I can crank this puppy out in a week, no sweat"), deep down we usually recognize risk well enough to feel scared. Mastering the complexities up front removes the fear and helps us work confidently and efficiently.

Conquer the Impossible

Firmware people are too often treated as the scum of the earth, because their development efforts tend to trail everyone else's. When the code can't be tested until the hardware is ready—and we know the hardware schedule is bound to slip—then the software, already starting late, will appear to doom the ship date.

Engineering is all about solving problems, yet sometimes we're immobilized like deer in headlights by the problems that litter our path. *We simply have to invent a solution to this dysfunctional cycle of starting firmware testing late because of unavailable hardware!*

And there are a lot of options.

One of the cheapest and most available tools around is the desktop PC. Use it! Here are a few ways to conquer the "I can't proceed because the hardware ain't ready" complaint.

- One compelling reason to use an embedded PC in non-cost-sensitive applications is that you can do much of the development on a standard PC. If your project permits, consider embedding a PC and plan on writing the code using standard desktop compilers and other tools.
- Write in C or C++. Cross-develop the code on a PC until hardware comes on line. It's amazing how much of the code you can get working on a different platform. Using a processor-specific timer or serial channel? Include conditional compilation switches to disable the target I/O and enable the PC's equivalent devices. One developer I know tests more than 95% of his code on the PC this way—and he's using a PIC processor, about as dissimilar from a PC as you can get.

- Regardless of processor, build an I/O board that contains your target-specific devices, such as A/Ds. There's an up-front time penalty incurred in creating the board; but the advantage is faster code delivery with more of the bugs wrung out. This step also helps prove the hardware design early—a benefit to everyone.

Summary

You'll never flatten the complexity/size curve unless you use every conceivable way to partition the code into independent chunks with no or few dependencies.

Some of these methods include the following:

- Partition by encapsulation
- Partition by adding CPUs
- Partition by using an RTOS (more in the next chapter)
- Partition by feature management and incremental development
- Finally, partition by top-down decomposition

CHAPTER **4**

Real Time Means Right Now!

We're taught to think of our code in the procedural domain: that of actions and effects. IF statements and control loops create a logical flow to implement algorithms and applications. There's a not-so-subtle bias in college toward viewing *correctness* as being nothing more than stringing the right statements together.

Yet embedded systems are the realm of real time, where getting the result on time is just as important as computing the correct answer.

A hard real-time task or system is one where an activity simply must be completed—always—by a specified deadline. The deadline may be a particular time or time interval, or may be the arrival of some event. Hard real-time tasks fail, by definition, if they miss such a deadline.

Notice that this definition makes no assumptions about the frequency or period of the tasks. A microsecond or a week—if missing the deadline induces failure, then the task has hard real-time requirements.

"Soft" real time, though, has a definition as weak as its name. By convention it's those class of systems that are not hard real time, though generally there is some sort of timeliness requirement. If missing a deadline won't compromise the integrity of the system, if generally getting the output in a timely manner is acceptable, then the application's real-time requirements are "soft." Sometimes soft real-time systems are those where multi-valued timeliness is acceptable: bad, better, and best responses are all within the scope of possible system operation.

Interrupts

Most embedded systems use at least one or two interrupting devices. Few designers manage to get their product to market without suffering metaphorical scars from battling interrupt service routines (ISRs). For some incomprehensible reason—perhaps because "real time" gets little more than lip service in academia—most of us leave college without the slightest idea of how to design, code, and debug these most important parts of our systems. Too many of us become experts at ISRs the same way we picked up the secrets of the birds and the bees—from quick conversations in the halls and on the streets with our pals. There's got to be a better way!

New developers rail against interrupts because they are difficult to understand. However, just as we all somehow shattered our parents' nerves and learned to drive a stick-shift, it just takes a bit of experience to become a certified "master of interrupts."

Before describing the "how," let's look at why interrupts are important and useful. Somehow peripherals have to tell the CPU that they require service. On a UART, perhaps a character arrived and is ready inside the device's buffer. Maybe a timer counted down and must let the processor know that an interval has elapsed.

Novice embedded programmers naturally lean toward polled communication. The code simply looks at each device from time to time, servicing the peripheral if needed. It's hard to think of a simpler scheme.

An interrupt-serviced device sends a signal to the processor's dedicated interrupt line. This causes the processor to screech to a stop and invoke the device's unique ISR, which takes care of the peripheral's needs. There's no question that setting up an ISR and associated control registers is a royal pain. Worse, the smallest mistake causes a major system crash that's hard to troubleshoot.

Why, then, not write polled code? The reasons are legion:

1. Polling consumes a lot of CPU horsepower. Whether the peripheral is ready for service or not, processor time—usually a lot of processor time—is spent endlessly asking "Do you need service yet?"
2. Polled code is generally an unstructured mess. Nearly every loop and long complex calculation has a call to the polling routines so that a device's needs never remain unserviced for long. ISRs, on the other hand, concentrate all of the code's involvement with each device into a single area. Your code is going to be a nightmare unless you encapsulate hardware-handling routines.

3. Polling leads to highly variable latency. If the code is busy handling something else (just doing a floating-point add on an 8-bit CPU might cost hundreds of microseconds), the device is ignored. Properly managed interrupts can result in predictable latencies of no more than a handful of microseconds.

Use an ISR pretty much any time a device can asynchronously require service. I say "pretty much" because there are exceptions. As we'll see, interrupts impose their own sometimes unacceptable latencies and overhead. I did a tape interface once, assuming the processor was fast enough to handle each incoming byte via an interrupt. Nope. Only polling worked. In fact, tuning the five instruction polling loops' speed ate up 3 weeks of development time.

Vectoring

Though interrupt schemes vary widely from processor to processor, most modern chips use a variation of *vectoring*. Peripherals, whether external to the chip or internal (such as on-board timers), assert the CPU's interrupt input.

The processor generally completes the current instruction and stores the processor's state (current program counter and possibly flag register) on the stack. The entire rationale behind ISRs is to accept, service, and return from the interrupt, all with no visible impact on the code. This is possible only if the hardware and software save the system's context before branching to the ISR.

It then *acknowledges* the interrupt, issuing a unique interrupt acknowledge cycle recognized by the interrupting hardware. During this cycle the device places an interrupt code on the data bus that tells the processor where to find the associated vector in memory.

Now the CPU interprets the vector and creates a pointer to the interrupt vector table, a set of ISR addresses stored in memory. It reads the address and branches to the ISR.

Once the ISR starts, you, the programmer, must preserve the CPU's context (such as saving registers, restoring them before exiting). The ISR does whatever it must, then returns with all registers intact to the normal program flow. The main-line application never knows that the interrupt occurred.

Figures 4-1 and 4-2 show two views of how an x86 processor handles an interrupt. When the interrupt request line goes high, the CPU completes the instruction it's executing (in this case at address 0100) and pushes the

56 THE ART OF DESIGNING EMBEDDED SYSTEMS

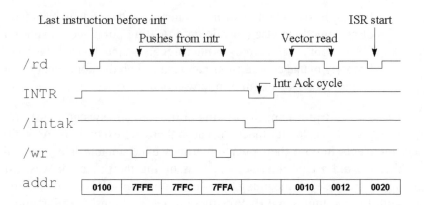

FIGURE 4-1 Logic analyzer view of an interrupt.

return address (two 16-bit words) and the contents of the flag register. The interrupt acknowledge cycle—wherein the CPU reads an interrupt number supplied by the peripheral—is unique, as there's no read pulse. Instead, intack going low tells the system that this cycle is unique.

x86 processors multiply the interrupt number by four (left shifted two bits) to create the address of the vector. A pair of 16-bit reads extracts the 32-bit ISR address.

Important points:

- The CPU chip's hardware, once it sees the interrupt request signal, does everything automatically, pushing the processor's state, reading the interrupt number, extracting a vector from memory, and starting the ISR.
- The interrupt number supplied by the peripheral during the acknowledge cycle might be hardwired into the device's brain, but

```
0100    NOP     Fetch       <-- INTR REQ asserted
7FFE    0102    Write       <-- Return address pushed
7FFC    0000    Write
7FFA    —       Write       <-- Flags pushed
xxxx    0010    INTA        <-- Vector inserted
0010    0020    Read        <-- ISR Address (low) read
0012    0000    Read        <-- ISR Address (high) read
0020    PUSH    Fetch       <-- ISR starts
```

FIGURE 4-2 Real-time trace view of an interrupt.

more commonly it's set up by the firmware. Forget to initialize the device and the system will crash as the device supplies a bogus number.
- Some peripherals and interrupt inputs will skip the acknowledge cycle because they have predetermined vector addresses.
- All CPUs let you disable interrupts via a specific instruction (DI, CLI, or something similar). Further, you can generally enable and disable interrupts from specific devices by appropriately setting bits in peripheral or interrupt control registers.
- Before invoking the ISR the hardware disables or reprioritizes interrupts. Unless your code explicitly reverses this, you'll see no more interrupts at this level.

At first glance the vectoring seems unnecessarily complicated. Its great advantage is support for many varied interrupt sources. Each device inserts a different vector; each vector invokes a different ISR. Your UART `Data_Ready` ISR is called independently of the UART `Transmit_Buffer_Full` routine. The vectoring scheme also limits pin counts, since it requires just one dedicated interrupt line.

Some CPUs sometimes directly invoke the ISR without vectoring. This greatly simplifies the code, but unless you add a lot of manual processing, it limits the number of interrupt sources a program can conveniently handle.

Interrupt Design Guidelines

While crummy code is just hard to debug, crummy ISRs are virtually undebuggable. The software community knows it's just as easy to write good code as it is to write bad. Give yourself a break and design hardware and software that eases the debugging process.

Poorly coded interrupt service routines are the bane of our industry. Most ISRs are hastily thrown together, tuned at debug time to work, then tossed in the "Oh my God, it works" pile and forgotten. A few simple rules can alleviate many of the common problems.

First, don't even consider writing a line of code for your new embedded system until you lay out an interrupt map (Figure 4-3). List each interrupt and give an English description of what the routine should do. Include your estimate of the interrupt's frequency. Figure the maximum, worst-case time available to service each. This is your guide: exceed this number, and the system stands no chance of functioning properly.

The map is a budget. It gives you an assessment of where interrupting time will be spent. Just as your own personal financial budget has a

	Latency	Max-time	Freq	Description
INT1		1000usec	1000usec	timer
INT2		100usec	100usec	send data
INT3		250usec	250usec	Serial data in
INT4		15usec	100usec	write tape
NMI	200usec	500usec	once!	System crash

FIGURE 4-3 An interrupt map.

degree of flexibility (spend too much on dinner this month and, assuming you don't abuse the credit cards, you'll have to reduce spending somewhere else). Like any budget, it's a condensed view of a profound reality whose parameters your system must meet. One number only is cast in stone: *there's only one second's worth of compute time per second to get everything done.* You can tune execution time of any ISR, but be sure there's enough time overall to handle every device.

Approximate the complexity of each ISR. Given the interrupt rate, with some idea of how long it'll take to service each, you can assign priorities (assuming your hardware includes some sort of interrupt controller). Give the highest priority to things that must be done in staggeringly short times to satisfy the hardware or the system's mission (such as to accept data coming in from a 1 Mb/sec source).

The cardinal rule of ISRs is to keep the handlers short. A long ISR simply reduces the odds you'll be able to handle all time-critical events in a timely fashion. If the interrupt starts something truly complex, have the ISR spawn off a task that can run independently. This is an area where an RTOS is a real asset, as task management requires nothing more than a call from the application code.

Short, of course, is measured in *time*, not in code size. Avoid loops. Avoid long complex instructions (repeating moves, hideous math, and the like). Think like an optimizing compiler: does this code *really* need to be in the ISR? Can you move it out of the ISR into some less critical section of code?

For example, if an interrupt source maintains a time-of-day clock, simply accept the interrupt and increment a counter. Then return. Let some other chunk of code—perhaps a non-real-time task spawned from the ISR—worry about converting counts to time and day of the week.

Ditto for command processing. I see lots of systems where an ISR receives a stream of serial data, queues it to RAM, and then executes commands or otherwise processes the data. Bad idea! The ISR should simply queue the data. If time is really pressing (i.e., you need real-time response to the data), consider using another task or ISR, one driven via a timer that interrupts at the rate you consider "real time," to process the queued data.

An analogous rule to keeping ISRs short is to keep them simple. Complex ISRs lead to debugging nightmares, especially when the tools may be somewhat less than adequate. Debugging ISRs with a simple BDM-like debugger is going to hurt—bad. Keep the code so trivial there's little chance of error.

An old rule of software design is to use one function (in this case the serial ISR) to do one thing. A real-time analogy is to do things *only when they need to get done*, not at some arbitrary rate.

Reenable interrupts as soon as practical in the ISR. Do the hardware-critical and non-reentrant things up front, then execute the interrupt enable instruction. Give other ISRs a fighting chance to do their thing.

Fill all of your unused interrupt vectors with a pointer to a null routine (Figure 4-4). During debug, *always* set a breakpoint on this routine. Any spurious interrupt, due to hardware problems or misprogrammed peripherals, will then stop the code cleanly and immediately, giving you a prayer of finding the problem in minutes instead of weeks.

Hardware Issues

Lousy hardware design is just as deadly as crummy software. Modern high-integration CPUs such as the 68332, 80186, and Z180 all include a wealth of internal peripherals—serial ports, timers, DMA controllers, etc. Interrupts from these sources pose no hardware design issues, since the chip vendors take care of this for you. All of these chips, though, do permit the use of external interrupt sources. There's trouble in them thar external interrupts!

```
Vect_table:
          dl    start_up          ; power up vector
          dl    null_isr          ; unused vector
          dl    null_isr          ; unused vector
          dl    timer_isr         ; main tick timer ISR
          dl    serial_in_isr     ; serial receive ISR
          dl    serial_out_isr    ; serial transmit ISR
          dl    null_isr          ; unused vector
          dl    null_isr          ; unused vector

null_isr:
          jmp   null_isr          ; spurious intr routine
                                  ; set BP here!
```

FIGURE 4-4 Fill unused vectors with a pointer to null_isr, and set a breakpoint there while debugging.

The biggest issue is the generation of the INTR signal itself. Don't simply pulse an interrupt input. Though some chips do permit edge-triggered inputs, the vast majority of them require you to assert and hold INTR until the processor issues an acknowledgment, such as from the interrupt ACK pin. Sometimes it's a signal to drop the vector on the bus; sometimes it's nothing more than "Hey, I got the interrupt—you can release INTR now."

As always, be wary of timing. A slight slip in asserting the vector can make the chip wander to an erroneous address. If the INTR must be externally synchronized to clock, do *exactly* what the spec sheet demands.

If your system handles a really fast stream of data, consider adding hardware to supplement the code. A data acquisition system I worked on accepted data at a 20-microsecond rate. Each generated an interrupt, causing the code to stop what it was doing, vector to the ISR, push registers like wild, and then reverse the process at the end of the sequence. If the system was busy servicing another request, it could miss the interrupt altogether.

A cheap 256-byte-deep FIFO chip eliminated all of the speed issues. The hardware filled the FIFO without CPU intervention. It generated an interrupt at the half-full point (modern FIFOs often have Empty, Half-Full, and Full bits), at which time the ISR sucked data from the FIFO until it was dry. During this process additional data might come along and be written to the FIFO, but this happened transparently to the code.

Most designs seem to connect FULL to the interrupt line. Conceptually simple, this results in the processor being interrupted only after the entire buffer is full. If a little extra latency causes a short delay before the CPU reads the FIFO, then an extra data byte arriving before the FIFO is read will be lost.

An alternative is EMPTY going not-true. A single byte arriving will cause the micro to read the FIFO. This has the advantage of keeping the FIFOs relatively empty, minimizing the chance of losing data. It also makes a big demand on CPU time, generating interrupts with practically every byte received.

Instead, connect HALF-FULL, if the signal exists on the FIFOs you've selected, to the interrupt line. HALF-FULL is a nice compromise, deferring processor cycles until a reasonable hunk of data is received, yet leaving free buffer space for more data during the ISR cycles.

Some processors do amazing things to service an interrupt, stacking addresses and vectoring indirectly all over memory. The ISR itself no doubt pushes lots of registers, perhaps also preserving other machine information. If the HALF-FULL line generates the interrupt, then you have the a priori information that lots of other data is already queued and waiting for processor time. Save overhead by making the ISR read the FIFOs until the EMPTY flag is set. You'll have to connect the EMPTY flag to a parallel port so the software can read it, but the increase in performance is well worth it.

In mission-critical systems it might also make sense to design a simple circuit that latches the combination of FULL and an incoming new data item. This overflow condition could be disastrous and should be signaled to the processor.

A few bucks invested in a FIFO may allow you to use a much slower, and cheaper, CPU. Total system cost is the only price issue in embedded design. If a $5 8-bit chip with a $6 FIFO does the work of a $20 16-bitter with double the RAM/ROM chips, it's foolish to not add the extra part.

Figure 4-5 shows the result of an Intel study of serial receive interrupts coming to a 386EX processor. At 530,000 baud—or around 53,000 characters per second—the CPU is almost completely loaded servicing interrupts.

Add a 16-byte FIFO and CPU loading declines to a mere 10%. That's a stunning performance improvement!

C or Assembly?

If you've followed my suggestions, you have a complete interrupt map with an estimated maximum execution time for the ISR. You're ready to start coding . . . right?

If the routine will be in assembly language, convert the time to a rough number of instructions. If an average instruction takes x microseconds (depending on clock rate, wait states, and the like), then it's easy to get this critical estimate of the code's allowable complexity.

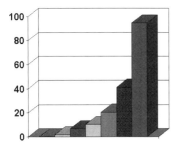
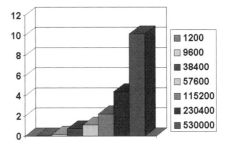

FIGURE 4-5   Baud rates versus CPU utilization. On the left, a conventional connection uses 90% of the CPU to service 530k baud inputs. On the right, with a FIFO the processor is 10% loaded at the same rate.

C is more problematic. In fact, there's no way to scientifically write an interrupt handler in C! You have no idea how long a line of C will take. You can't even develop an estimate as each line's time varies wildly. A string compare may result in a runtime library call with totally unpredictable results. A FOR loop may require a few simple integer comparisons or a vast amount of processing overhead.

And so, we write our C functions in a fuzz of ignorance, having no concept of execution times until we actually run the code. If it's too slow, well, just change something and try again!

I'm not recommending that ISRs not be coded in C. Rather, this is more of a rant against the current state of compiler technology. Years ago assemblers often produced t-state counts on the listing files, so you could easily figure how long a routine ran. Why don't compilers do the same for us? Though there are lots of variables (that string compare will take a varying amount of time depending on the data supplied to it), certainly many C operations will give deterministic results. It's time to create a feedback loop that tells us the cost, in time and bytes, for each line of code we write, before burning ROMs and starting test.

Until compilers improve, use C if possible, but look at the code generated for a typical routine. Any call to a runtime routine should be immediately suspect, as that routine may be slow or non-reentrant, two deadly sins for ISRs. Look at the processing overhead—how much pushing and popping takes place? Does the compiler spend a lot of time manipulating the stack frame? You may find one compiler pitifully slow at interrupt handling. Either try another, or switch to assembly.

> Despite all of the hype you'll read in magazine ads about how vendors understand the plight of the embedded developer, the plain truth is that the compiler vendors all uniformly miss the boat. Modern C and C++ compilers are poorly implemented in that they give us no feedback about the real-time nature of the code they're producing.
>
> The way we write performance-bound C code is truly astounding. Write some code, compile and run it . . . and if it's not fast enough, change something—anything—and try again. The compiler has so distanced us from the real-time nature of the code that we're left to make random changes in a desperate attempt to get the tool to produce faster code.
>
> A much more reasonable approach would be to get listings from the compiler with typical per-statement execution times. An ideal listing might resemble

```
250-275 nsec      for(i=0; i<count; ++i)
508-580 nsec      {if (start_count !=
                   end_count)
250 nsec          end_point=head;
                  }
```

where a range of values cover possible differences in execution paths depending on how the statement operates (for example, if the FOR statement iterates or terminates).

To get actual times, of course, the compiler needs to know a lot about our system, including clock rates and wait states. Another option is to display T states, or even just number of instructions executed (since that would give us at least *some* sort of view of the code's performance in the time domain).

Vendors tell me that cache, pipelines, and prefetchers make modeling code performance too difficult. I disagree. Most small embedded CPUs don't have these features, and of them, only cache is truly tough to model.

Please, Mr. Compiler Vendor, give us some sort of indication about the sort of performance we can expect! Give us a clue about how long a runtime routine or floating-point operation takes.

A friend told me how his DOD project uses an antique language called CMSP. The compiler is so buggy they have to look for bugs in the assembly listing after each and every compile—and then make a more or less random change and recompile, hoping to lure the tool into creating correct code. I laughed until I realized that's exactly the situation we're in when using a high-quality C compiler in performance-bound applications.

Be especially wary of using complex data structures in ISRs. Watch what the compiler generates. You may gain an enormous amount of performance by sizing an array at an even power of 2, perhaps wasting some memory, but avoiding the need for the compiler to generate complicated and slow indexing code.

An old software adage recommends coding for functionality first, and speed second. Since 80% of the speed problems are usually in 20% of the code, it makes sense to get the system working and then determine where the bottlenecks are. Unfortunately, real-time systems by their nature usually don't work at all if things are slow. You've often *got* to code for speed up front.

If the interrupts are coming fast—a term that is purposely vague and qualitative, measured by experience and gut feel—then I usually just take the plunge and code the ISR in assembly. Why cripple the entire system because of a little bit of interrupt code? If you've broken the ISRs into small chunks, so the real-time part is small, then little assembly will be needed. Code the slower ISRs in C.

Debugging INT/INTA Cycles

Lots of things can and will go wrong long before your ISR has a chance to exhibit buggy behavior. Remember that most processors service an interrupt with the following steps:

1. The device hardware generates the interrupt pulse.
2. The interrupt controller (if any) prioritizes multiple simultaneous requests and issues a single interrupt to the processor.
3. The CPU responds with an interrupt acknowledge cycle.
4. The controller drops an interrupt vector on the databus.
5. The CPU reads the vector and computes the address of the user-stored vector in memory. It then fetches this value.
6. The CPU pushes the current context, disables interrupts, and jumps to the ISR.

Interrupts from internal peripherals (those on the CPU itself) usually won't generate an external interrupt acknowledge cycle. The vectoring is handled internally and invisibly to the wary programmer, tools in hand, trying to discover his system's faults.

A generation of structured programming advocates has caused many of us to completely design the system and write all of the code before debugging. Though this is certainly a nice goal, it's a mistake for the low-level drivers in embedded systems. I believe in an early wrestling match with the system's hardware. Connect an emulator and exercise the I/O ports. They never behave quite as you expected. Bits might be inverted or transposed, or maybe there are a dozen complex configuration registers that need to be set up. Work with your system, understand its quirks, and develop notes about how to drive each I/O device. Use these notes to write your code.

Similarly, start prototyping your interrupt handlers with a hollow shell of an ISR. You've got to get a lot of things *right* just to get the ISR to start. Don't worry about what the handler should do until you have it at least being called properly.

Set a breakpoint on the ISR. If your shell ISR never gets called, and the system doesn't crash and burn, most likely the interrupt never makes it to the CPU. If you were clever enough to fill the vector table's unused entries with pointers to a null routine, watch for a breakpoint on that function. You may have misprogrammed the table entry or the interrupt controller, which would then supply a wrong vector to the CPU.

If the program vectors to the wrong address, then use a logic analyzer or emulator's trace to watch how the CPU services the interrupt. Trigger collection on the interrupt itself, or on any read from the vector table in RAM. You should see the interrupt controller drop a vector on the bus. Is it the right one? If not, perhaps the interrupt controller is misprogrammed.

Within a few instructions (if interrupts are on) look for the read from the vector table. Does it access the right table address? If not, and if the vector was correct, then either you're looking at the wrong system interrupt, or there's a timing problem in the interrupt acknowledge cycle. Break out the logic analyzer and check this carefully.

Hit the databooks and check the format of the table's entries. On an x86-style processor, four bytes represent the ISR's offset and segment address. If these are in the wrong order—and they often are—there's no chance your ISR will execute.

Frustratingly often the vector is fine; the interrupt just does not occur. Depending on the processor and peripheral mix, only a handful of things could be wrong:

- Did you enable interrupts in the main routine? Without an EI instruction, no interrupt will ever occur. One way of detecting this is to sense the CPU's INTR input pin. If it's asserted all of the time, then generally the chip has all interrupts disabled.
- Does your I/O device generate an interrupt? It's easy to check this with external peripherals.
- Have you programmed the device to allow interrupt generation? Most CPUs with internal peripherals allow you to selectively disable each device's interrupt generation; quite often you can even disable parts of this (such as allow interrupts on "received data" but not on "data transmitted").

Modern peripherals are often incredibly complex. Motorola's TPU, for example, has an entire book dedicated to its use. Set one bit in one register to the wrong value, and it won't generate the interrupt you are looking for.

It's not uncommon to see an interrupt work perfectly once, and then never work again. The only general advice is to be sure your ISR reenables interrupts before returning. Then look into the details of your processor and peripherals.

Some, such as the Z80, have an external interrupt daisy chain that serves as a priority encoder. Look at these lines with a scope. If you see the daisy chain set to a zero, it's a sure indication that one device did not see the end-of-interrupt sequence. On the Z80 and Z180 processors this is provided by executing the RETI instruction. Use a normal return instruction by mistake and you'll never get another interrupt.

Intel's x86 family is often used with an 8259 interrupt controller. Some of the embedded CPUs in this family have 8259-like controllers built into the processor. If you forget to issue an EOI (end of interrupt) command to the 8259 when the ISR is complete, you'll get that one interrupt only.

You may need to service the peripherals as well before another interrupt comes along. Depending on the part, you may have to read registers in the peripheral to clear the interrupt condition. UARTs and timers usually require this. Some have peculiar requirements for clearing the interrupt condition, so be sure to dig deeply into the databook.

Finding Missing Interrupts

A device that parses a stream of incoming characters will probably crash very obviously if the code misses an interrupt or two. One that counts interrupts from an encoder to measure position may only exhibit small precision errors, a tough thing to find and troubleshoot.

Having worked on a number of systems using encoders as position sensors, I've developed a few tricks over the years to find these missing pulses.

You can build a little circuit using a single up/down counter that counts every interrupt and that decrements the count on each interrupt acknowledge. If the counter always shows a value of zero or one, everything is fine.

Most engineering labs have counters—test equipment that just accumulates pulse counts. I have a scope that includes a counter. Use two of these, one on the interrupt pin and another on the interrupt acknowledge pin. The counts should always be the same.

You can build a counter by instrumenting the ISR to increment a variable each time it starts. Either show this value on a display, or probe the variable using your debugger.

If you know the maximum interrupt rate, use a performance analyzer to measure the maximum time in the ISR. If this exceeds the fastest interrupts, there's very likely a latent problem waiting to pounce.

Most of these sorts of difficulties stem from slow ISRs, or from code that leaves interrupts off for too long. Be wary of any code that executes a disable-interrupt instruction. There's rarely a good reason for it; this is usually an indication of sloppy software.

It's rather difficult to find a chunk of code that leaves interrupts off. The ancient 8080 had a wonderful pin that showed interrupt state all of the time. It was easy to watch this on the scope and look for interrupts that came during that period. Now, having advanced so far, we have no such easy troubleshooting aids. About the best one can do is watch the INTR pin. If it stays asserted for long periods of time, and if it's properly designed (i.e., stays asserted until INTA), then interrupts are certainly off.

One design rule of thumb will help minimize missing interrupts: reenable interrupts in ISRs at the earliest safe spot.

Reentrancy Problems

Well-designed interrupt handlers are largely reentrant. Reentrant functions—a.k.a. "pure code"—are often falsely thought to be any code that does not modify itself. Too many programmers feel that if they simply avoid self-modifying code, their routines are guaranteed to be reentrant, and thus interrupt-safe. Nothing could be further from the truth.

A function is reentrant if, while it is being executed, it can be reinvoked by itself, or by any other routine.

Suppose your main-line routine and the ISRs are all coded in C. The compiler will certainly invoke runtime functions to support floating-point math, I/O, string manipulations, etc. If the runtime package is only partially reentrant, then your ISRs may very well corrupt the execution of the main line code. This problem is common, but is virtually impossible to troubleshoot, since symptoms result only occasionally and erratically. There's nothing more ulcer-inducing than isolating a bug that manifests itself only occasionally, and with totally different characteristics each time.

Sometimes we're tempted to cheat and write a nearly pure routine. If your ISR merely increments a global 32-bit value, maybe to maintain time, it would seem legal to produce code that does nothing more than a quick and dirty increment. Beware! Especially when writing code on an 8- or 16-bit processor, remember that the C compiler will surely generate several instructions to do the deed. On a 186, the construct ++j might produce

```
mov     ax,[j]
add     ax,1         ; increment low part of j
mov     [j],ax
```

```
mov     ax,[j+1]
adc     ax,0            ; prop carry to high part of j
mov     [j+1],ax
```

An interrupt in the middle of this code will leave j just partially changed; if the ISR is reincarnated with j in transition, its value will surely be corrupt. Or, if other routines use the variable, the ISR may change its value at the same time other code tries to make sensible use of it.

The first solution is to avoid global variables! Globals are an abomination, a sure source of problems in any system, and an utter nightmare in real-time code. Never, ever pass data between routines in globals unless the following three conditions are fulfilled:

- Reentrancy issues are dealt with via some method, such as disabling interrupts around their use—though I do not recommend disabling interrupts cavalierly, since that affects latency.
- The globals are absolutely needed because of a clear performance issue. Most alternatives do impose some penalty in execution time.
- The global use is limited and well documented.

Inside of an ISR, be wary of any variable declared as a static. Though statics have their uses, the ISR that reenables interrupts, and then is interrupted before it completes, will destroy any statics declared within.

In 1997, on a dare, I examined firmware embedded in 23 completed products, all of which were shipping to customers. Every one had this particular problem! Interestingly, the developers of 70% of the projects admitted to infrequent, unexplainable crashes or other odd behavior. One frustrated engineer revealed that his product burped almost hourly, a symptom "corrected" (perhaps "masked" would be a better term) by adding a very robust watchdog timer circuit. This particularly bad system, which had the reentrancy problem inside an ISR, also had the fastest interrupt rate of any of the products examined.

This suggests using a stress test to reveal latent reentrancy defects. Crank up the interrupt rates! If the timer comes once per second, try driving it every millisecond and see how the system responds. Assuming performance issues don't crash the code, this simple test often shows a horde of hidden flaws.

Even the perfectly coded reentrant ISR leads to problems. If such a routine runs so slowly that interrupts keep giving birth to additional copies of it, eventually the stack will fill. Once the stack bangs into your variables, the program is on its way to oblivion. You *must* ensure that the average interrupt rate is such that the routine will return more often than it is invoked. Again, use the stress test!

Avoid NMI

Reserve NMI—the non-maskable interrupt—for a catastrophe such as the apocalypse. Power-fail, system shutdown, and imminent disaster are all good things to monitor with NMI. Timer or UART interrupts are not.

When I see an embedded system with the timer tied to NMI, I know, for sure, that the developers found themselves missing interrupts. NMI may alleviate the symptoms, but only masks deeper problems in the code that *must* be cured.

NMI will break even well-coded interrupt handlers, since most ISRs are non reentrant during the first few lines of code where the hardware is serviced. NMI will thwart your stack-management efforts as well.

If you're using NMI, watch out for electrical noise! NMI is usually an edge-triggered signal. Any bit of noise or glitching will cause perhaps hundreds of interrupts. Since it cannot be masked, you'll almost certainly cause a reentrancy problem. I make it a practice to always properly terminate the CPU's NMI input via an appropriate resistor network.

NMI mixes poorly with most tools. Debugging any ISR—NMI or otherwise—is exasperating at best. Few tools do well with single stepping and setting breakpoints inside of the ISR.

Breakpoint Problems

Using any sort of debugging tool, suppose you set a breakpoint where the ISR starts, and then start single stepping through the code. All is well, since by definition interrupts are off when the routine starts. Soon, though, you'll step over an EI instruction or its like. Suddenly, all hell breaks lose.

A regularly occurring interrupt such as a timer tick comes along steadily, perhaps dozens or hundreds of times per second. Debugging at human speeds means the ISR will start over while you're working on a previous instantiation. Pressing the "single step" button might make the ISR start, but then itself be interrupted and restarted, with the final stop due to your high-level debug command coming from this second incarnation.

Oddly, the code seems to execute backwards. Consider the case of setting two breakpoints—the first at the start of the ISR and the second much later into the routine. Run to the first breakpoint, stop, and then resume execution. The code may very well stop at the same point, the same first breakpoint, without ever going to the second. Again, this is simply due to the human-speed debugging that gives interrupting hardware a chance to issue yet another request while the code's stopped at the breakpoint.

In the case of NMI, though, disaster strikes immediately, since there is no interrupt-safe state. The NMI is free to reoccur at any time, even in the most critical non-reentrant parts of the code, wreaking havoc and despair.

A lot of applications now just can't survive the problems inherent in using breakpoints. After all, stopping the code stops everything; your entire system shuts down. If your code controls a moving robot arm, for example, and you stop the code as the arm starts moving, it will keeping going and going and going . . . until something breaks or a limit switch is actuated. Years ago I worked on a 14-ton steel gauge; a Z80 controlled the motion of this monster on railroad tracks. Hit a breakpoint and the system ran off the end of the tracks!

Datacomm is another problem area. Stop the code via a breakpoint, with data packets still streaming in, and there's a good chance the receiving device will time out and start transmitting retry requests.

Though breakpoints are truly wonderful debugging aids, they are like Heisenberg's uncertainty principle: the act of looking at the system changes it. You can cheat Heisenberg—at least in debugging embedded code!—by using real-time trace, a feature available on all emulators and some smart logic analyzers.

Trace collects the execution stream of the code in real time, without slowing or altering the flow. It's a completely nonintrusive way of viewing what happens.

Trace changes the philosophy of debugging. No longer does one stop the code, examine various registers and variables, and then timidly step along. With trace your program is running at full tilt, a breakneck pace that trace does nothing to alter. You capture program flow, and then examine what happened, essentially looking into the past as the code continues on (Figure 4-6).

Trace shows only what happens on the bus. You can view neither registers nor variables unless an instruction reads or writes them to memory. Worse, C's stack-based design often makes it impossible to view variables that were captured. You may see the transactions (pushes and pops), but the tool may display neither the variable name nor the data in its native type.

With millions of instructions every second, it's clearly impossible to capture your program's entire execution stream. Nor is it desirable, as a trace buffer a hundred million frames deep is simply too much data to plow through. Pick an emulator that offers flexible triggers—breakpoint-like resources that start and stop trace collection.

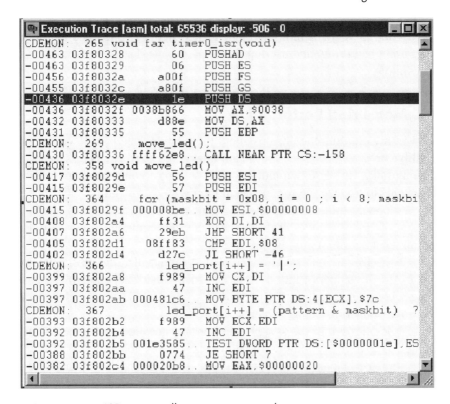

FIGURE 4-6 ISR trace collection on an emulator.

Are the triggers a pain to set up? Most emulators offer special menus with dozens of trigger configuration options. Although this is essential for finding the most obscure bugs, it is just too much work for the usual debugging scenario, where you simply want to start collection when source module line 124 executes. Simple triggers should be as convenient as breakpoints, set perhaps via a right mouse click.

The moral is: trace is the right debugging tool, but keep ISRs simple. Minimize their complexity to maximize their debuggability.

Easy ISR Debugging

What's the fastest way to debug an ISR?
Don't.
If your ISR is only 10 or 20 lines of code, debug by inspection. Don't fire up all kinds of complex and unpredictable tools.

Keep the handler simple and short. If it fails to operate correctly, a few minutes reading the code will usually uncover the problem.

After 25 years of building embedded systems I've learned that long ISRs are a bad thing, and a symptom of poor code. Keep 'em short, keep 'em simple.

Measuring Performance

In my opinion, the debates about the relative speeds of C versus assembly, or C versus C++, are meaningless. All performance issues are nothing but a smokescreen unless you're willing to take qualitative measurements to replace the fog of speculation with the insight of facts.

Amateurs moan and speculate about performance, making random stabs at optimizing code. Professionals take measurements, only then deciding what action, if any, is appropriate.

If the ISR is not fast enough, your system will fail. Unfortunately, few of the developers I talk to have any idea what "fast enough" means. Unless you generate the interrupt map I've discussed, only random luck will save you from speed problems.

When designing the system, answer two questions: how fast is fast enough? How will you know if you've reached this goal?

Some people are born lucky. Not me. I've learned that nature is perverse and will get me if it can. Call it high-tech paranoia. Plan for problems, and develop solutions for those problems before they occur. Assume each ISR will be too slow, and plan accordingly.

A performance analyzer will instantly show the minimum, maximum, and average execution time required by your code, including your ISRs (Figure 4-7). There's no better tool for finding real-time speed issues.

Guesstimating Performance

In 1967 Keuffel & Esser (the greatest of the slide rule companies) commissioned a study of the future. They predicted that by 2067 we'd see three-dimensional TVs and cities covered by majestic domes. The study somehow missed the demise of the slide rule (their main product) within 5 years.

Our need to compute, to routinely deal with numbers, led to the invention of dozens of clever tools, from the abacus to logarithm tables to the slide rule. All worked in concert with the user's brain, in an iterative, back-and-forth process that only slowly produced answers.

Now even grade-school children routinely use graphing calculators. The device assumes the entire job of computation and sometimes even data analysis. What a marvel of engineering! Powered by nothing more than a

Real Time Means Right Now! **73**

Function Name	Min	Max	Avg	Total	Count	%
main()	5.952 uS	5.952 uS	90.005 mS	90.005 mS	00C01	40.00
vBank000(	374.976 uS	725.152 uS	550.064 uS	1.100 mS	00C02	1.45
vBank001(	720.192 uS	1.421 mS	1.070 mS	2.141 mS	00C02	1.89
vBank002(	1.080 mS	2.132 mS	1.606 mS	3.212 mS	00C02	1.33
vBank003(	1.421 mS	2.822 mS	2.121 mS	4.243 mS	00C02	1.76
vBank004(	1.782 mS	3.534 mS	2.658 mS	5.315 mS	00C02	2.21
vBank005(	2.121 mS	4.223 mS	3.172 mS	6.344 mS	00C02	2.64
vBank006(	2.482 mS	4.933 mS	3.708 mS	7.415 mS	00C02	3.08
vBank007(	2.822 mS	5.624 mS	4.223 mS	8.446 mS	00C02	3.51
vBank008(	3.182 mS	6.335 mS	4.759 mS	9.517 mS	00C02	3.96
vBank009(	19.840 uS	24.800 uS	24.800 uS	49.600 uS	00C02	1.02
vBank010(	370.016 uS	720.192 uS	545.104 uS	1.090 mS	00C02	1.45
vBank011(	731.104 uS	1.430 mS	1.081 mS	2.162 mS	00C02	1.90
vBank012(	1.070 mS	2.121 mS	1.596 mS	3.191 mS	00C02	1.32
vBank013(	1.431 mS	2.832 mS	2.132 mS	4.264 mS	00C02	1.77
vBank014(	1.771 mS	3.523 mS	2.647 mS	5.293 mS	00C02	2.20
vBank015(	2.132 mS	4.234 mS	3.183 mS	6.366 mS	00C02	2.65
vBank016(	2.471 mS	4.923 mS	3.697 mS	7.394 mS	00C02	3.07
vBank017(	2.832 mS	5.635 mS	4.233 uS	8.467 mS	00C02	3.52

FIGURE 4-7   A performance analyzer's output.

stream of photons, pocket-sized, and costing virtually nothing, our electronic creations give us astonishing new capabilities.

Those of us who spend our working lives parked in front of computers have even more powerful computational tools. The spreadsheet is a multidimensional version of the hand calculator, manipulating thousands of formulas and numbers with a single keystroke. Excel is one of my favorite engineering tools. It lets me model weird systems without writing a line of code, and tune the model almost graphically. Computational tools have evolved to the point where we no longer struggle with numbers; instead, we ask complex "what-if" questions.

Network computing lets us share data. We pass spreadsheets and documents among co-workers with reckless abandon. In my experience, big, widely shared spreadsheets are usually incorrect. Someone injects a row or column, forgetting to adjust a summation or other formula. The data at the end is so complex, based on so many intermediate steps, that it's hard to see if it's right or wrong . . . so we assume it's right. This is the dark side of a spreadsheet: no other tool can make so many incorrect calculations so fast.

Mechanical engineers now use finite element analysis to predict the behavior of complex structures under various stresses. The computer models a spacecraft vibrating as it is boosted to orbit, giving the designers insight into its strength without the need to run expensive tests on shakers. Yet, finite element analysis is so complex, with millions of interrelated calculations! How do they convince themselves that a subtle error isn't lurking in the model? As with subtle errors hidden in large spreadsheets, the complexity of the calculations removes the element of "feel." Is that complex carbon-fiber structure strong enough when excited at 20 Hz? Only the computer knows for sure.

The modern history of engineering is one of increasing abstraction from the problem at hand. The C language insulates us from the tedium of assembly, which itself removes us from machine code. Digital ICs protect us from the very real analog behavior of each of the millions of transistors encapsulated in the chip. When we embed an operating system into a product, we're given a wealth of services we can use without really understanding the how and why of their operation.

Increasing abstraction is both inevitable and necessary. An example is the move to object-oriented programming, and more importantly, software reuse, which will—someday—lead to "software ICs" whose operation is as mysterious as today's giant LSI devices, yet that elegantly and cheaply solve some problem.

But, abstraction comes at a price. In too many cases we're losing the "feel" of the problem. Engineering has always been about building things, in the most literal of contexts. Building, touching, and experiencing failure are the tactile lessons that burn themselves into the wiring of our brains. When we delve deeply into how and why things work, when we get burned by a hot resistor, when we've had a tantalum capacitor installed backwards explode in our face, when a CMOS device fails from excessive undershoot on an input, we develop our own rules of thumb that give us a new understanding of electronics. Book learning tells us what we need to know. Handling components and circuits builds a powerful subconscious knowledge of electronics.

A friend who earns his keep as a consultant sometimes has to admit that a proposed solution looks good on paper, but just does not feel right. Somehow we synthesize our experience into an emotional reaction as powerful and immediate as any other feeling. I've learned to trust that initial impression, and to use that bit of nausea as a warning that something is not quite right. The ground plane on that PCB just doesn't look heavy enough. The capacitors seem a long way from the chips. That sure seems like a long cable for those fast signals. Gee, there's a lot of ringing on that node.

Practical experience has always been an engineer's stock-in-trade. We learn from our successes and our failures. This is nothing new. According to *Cathedral, Forge and Waterwheel* (Frances and Joseph Gies, 1994, HarperCollins, New York), in the Middle Ages "Engineers had some command of geometry and arithmetic. What they lacked was engineering theory, in place of which they employed their own experience, that of their colleagues, and rule of thumb."

The flip side of a "feel" for a problem is an ability to combine that feeling with basic arithmetic skills to very quickly create a first approximation to a solution, something often called "guesstimating." This wonderful word combines "guess"—based on our engineering feel for a problem—and "estimate"—a partial analytical solution.

Guesstimates are what keep us honest: "200,000 bits per second seems kind of fast for an 8-bit micro to process" (this is the guess part); "Why, that's 1/200,000 or 5 microseconds per bit" (the estimate part). Maybe there's a compelling reason why this guesstimate is incorrect, but it flags an area that needs study.

In 1995 an Australian woman swam the 110 miles from Havana to Key West in 24 hours. Public Radio reported this information in breathless excitement, while I was left baffled. My guesstimate said this is unlikely. That's a 4.5 MPH average, a pace that's hard to beat even with a brisk walk, yet the she maintained this for a solid 24 hours.

Maybe swimmers are speedier than I'd think. Perhaps the Gulf Stream spun off a huge gyre, a rotating current that gave her a remarkable boost in the right direction. I'm left puzzled, as the data fails my guesstimating sense of reasonableness. And so, though our sense of "feel" can and should serve as a measure against which we can evaluate the mounds of data tossed our way each day, it is imperfect at best.

The art of "guesstimating" was once the engineer's most basic tool. Old engineers love to point to the demise of the slide rule as the culprit. "Kids these days," they grumble. Slide rules forced one to estimate the solution to every problem. The slide rule did force us to have an easy familiarity with numbers and with making coarse but rapid mental calculations.

We forget, though, just how hard we had to work to get anything done! Nothing beats modern technology for number crunching, and I'd never go back. Remember that the slide rule *forced* us to estimate all answers; the calculator merely *allows* us to accept any answer as gospel without doing a quick mental check.

We need to grapple with the size of things, every day and in every avenue. A million times a million is, well, 10^{12}. The gigahertz is a period of one nanosecond. A speed of 4.5 miles per hour seems high for a swimmer. It's unlikely your interrupt service routine will complete in 2 microseconds.

We're building astonishing new products, the simplest of which have hundreds of functions requiring millions of transistors. Without our amazing tools and components, those things that abstract us from the worries of biasing each individual transistor, we'd never be able to get our work done.

Though the abstraction distances us from how things work, it enables us to make things work in new and wondrous ways.

The art of guesstimating fails when we can't or don't understand the system. Perhaps in the future we'll need computer-aided guesstimating tools, programs that are better than feeble humans at understanding vast interlocked systems. Perhaps this will be a good thing. Maybe, like double-entry bookkeeping, a computerized guesstimator will at least allow a cross-check on our designs.

When I was a nerdy kid in the 1960s, various mentors steered me to vacuum tubes long before I ever understood semiconductors. A tube is wonderfully easy to understand. Sometimes you can quite literally see the blue glow of electrons splashing off the plate onto the glass. The warm glow of the filaments, the visible mesh of the control grids, always conjured a crystal-clear mental image of what was going on.

A 100,000-gate ASIC is neither warm nor clear. There's no emotional link between its operation and your understanding of it. It's a platonic relationship at best.

So, what's an embedded engineer to do? How can we reestablish this "feel" for our creations, this gut-level understanding of what works and what doesn't?

The first part of learning to guesstimate is to gain an intimate understanding of how things work. We should encourage kids to play with technology and science. Help them get their hands greasy. It matters little if they work on cars, electronics, or in the sciences. Nurture that odd human attribute that couples doing with learning.

The second part of guesstimation is a quick familiarity with math. Question engineers (and your kids) deeply about things. "Where did that number come from?" "Do you believe it . . . and why?"

Work on your engineer's understanding of orders of magnitude. It's astonishing how hard some people work to convert frequency to period, yet this is the most common calculation we do in computer design. If you know that a microsecond is a megahertz, a millisecond is 1000 Hz, you'll never spend more than a second getting a first-approximation conversion.

The third ingredient is to constantly question everything. As the bumper sticker says, "Question authority." As soon as the local expert backs up his opinion with numbers, run a quick mental check. He's probably wrong.

In *To Engineer Is Human* (1982, Random House, New York), author Henry Petroski says, "Magnitudes come from a *feel* for the problem, and do not come automatically from machines or calculating contrivances." Well put, and food for thought for all of us.

A simple CPU has very predictable timing. Add a prefetcher or pipeline and timing gets fuzzier, but still is easy to figure within 10 or 20%. Cache is the wildcard, and as cache size increases, determinism diminishes. Thankfully, today few small embedded CPUs have even the smallest amount of cache.

Your first weapon in the performance arsenal is developing an understanding of the target processor. What can it do in one microsecond? One instruction? Five? Some developers use very, very slow clocks when not much has to happen—one outfit I know runs the CPU (in a spacecraft) at 8 kHz until real speed is needed. At 8 kHz they get maybe 1000 instructions per second. Even small loops become a serious problem. Understanding the physics—a perhaps fuzzy knowledge of just what the CPU can do at this clock rate—means the big decisions are easy to make.

Estimation is one of engineering's most important tools. Do you think the architect designing a house does a finite element analysis to figure the size of the joists? No! He refers to a manual of standards. A 15-foot unsupported span typically uses joists of a certain size. These estimates, backed up with practical experience, ensure that a design, while perhaps not optimum, is adequate.

We do the same in hardware engineering. Electrons travel at about one or two feet per nanosecond, depending on the conductor. It's hard to make high-frequency first harmonic crystals, so use a higher order harmonic. Very small PCB tracks are difficult to manufacture reliably. All of these are ingredients of the "practice" of the art of hardware design. None of these are tremendously accurate: you can, after all, create one-mil tracks on a board for a ton of money. The exact parameters are fuzzy, but the general guidelines are indeed correct.

So, too, for software engineering. We need to develop a sense of the art. A 68HC16, at 16 MHz, runs so many instructions per second (plus or minus). With this particular compiler you can expect (more or less) this sort of performance under these conditions.

Data, even fuzzy data, lets us bound our decisions, greatly improving the chances of success. The alternative is to spend months and years generating a mathematically precise solution—which we won't do—or to burn incense and pray . . . the usual approach.

Experiment. Run portions of the code. Use a stopwatch—metaphorical or otherwise—to see how it executes. Buy a performance analyzer or simply instrument sections of the firmware to understand the code's performance.

The first time you do this you'll think, "This is so cool," and you'll walk away with a clear number: xxx microseconds for this routine. With time you'll develop a sense of speed. "You know, integer compares are pretty damn fast on this system." Later—as you develop a sense of the art—you'll be able to bound things. "Nah, there's no way that loop can complete in 50 microseconds."

This is called experience, something that we all too often acquire haphazardly. We plan our financial future, we work daily with our kids on their homework, even remember to service the lawnmower at the beginning of the season, yet neglect to proactively improve our abilities at work.

Experience comes from exposure to problems and from learning from them. A fast, useful sort of performance expertise comes from extrapolating from a current product to the next. Most of us work for a company that generally sells a series of similar products. When it's time to design a new one, we draw from the experience of the last, and from the code and design base. Building version 2.0 of a widget? Surely you'll use algorithms and ideas from 1.0. Use 1.0 as a testbed. Gather performance data by instrumenting the code.

Always close the feedback loop! When any project is complete, spend a day learning about what you did. Measure the performance of the system to see just how accurate your processor utilization estimates were. The results are always interesting and sometimes terrifying. If, as is often the case, the numbers bear little resemblance to the original goals, then figure out what happened, and use this information to improve your estimating ability. Without feedback, you work forever in the dark. Strive to learn from your successes as well as your failures.

Track your system's performance all during the project's development, so you're not presented with a disaster two weeks before the scheduled delivery. It's not a bad idea to assign CPU utilization specifications to major routines during overall design, and then track these targets as you do the schedule. Avoid surprises with careful planning.

A lot of projects eventually get into trouble by overloading the processor. This is always discovered late in the development, during debugging or final integration, when the cost of correcting the problem is at the maximum. Then a mad scramble to remove machine cycles begins.

We all know the old adage that 80% of the processor burden lies in 20% of the code. It's important to find and optimize that 20%, not some other section that will have little impact on the system's overall performance. Nothing is worse than spending a week optimizing the wrong routine!

If you understand the design, if you have a sense of the CPU, you'll know where that 20% of the code is before you write a line. Knowledge is power.

Learn about your hardware. Pure software types often have no idea that the CPU is actively working against them. I talked to an engineer lately who was moaning about how slow his new 386EX-based instrument runs. He didn't know that the 386EX starts with 31 wait states and so had never reprogrammed it to a saner value.

A Poor Man's Performance Analyzer

Do keep in tune with the embedded tool industry's wide range of performance-analyzing devices. But don't fail to take detailed measurements just because such a tool is not available. An oscilloscope coupled to a few spare output bits can be a very effective and cheap performance analyzer.

Whether you're working on an 8-bit microcontroller or a 32-bit VME-based system, always dedicate one or two parallel I/O bits to debugging. That is, *have the hardware designers include a couple of output bits just for software debugging purposes.* The cost is vanishingly small; the benefits often profound.

Suppose you'd like to know an ISR's (or any other sort of routine's) precise execution time. Near the beginning of the routine set a debug output bit high; just before exiting return the bit to a zero. For example:

```
ISR_entry:
    push all registers
    set output bit high
    service interrupt
    reset output bit
    pop registers
    return
```

Put one scope probe on the bit. You'll see a pattern that might resemble that in Figure 4-8. The ISR is executing when the signal is high.

In this example we see two invocations of the ISR. The first time (note that the time base setting is 2 msec/division), the routine runs for a bit over 3 msec. Next time (presumably the routine includes conditional code), it runs for under 1 msec.

We also clearly see a 14-msec period between executions. If these two samples are indicative of the system's typical operation, the total CPU overhead dedicated to this one interrupt is (3 msec+1 msec)/14 msec, or 29%.

Crank up the scope's time base and you can measure the ISR's execution time to any desired level of precision.

80 THE ART OF DESIGNING EMBEDDED SYSTEMS

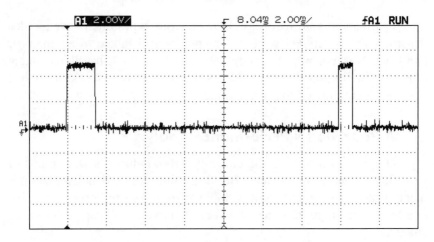

FIGURE 4-8 Measuring an ISR's execution time.

When I see a 29% CPU loading for a single ISR, I immediately wonder why the ISR takes so much time. It violates my commonsense, guesstimating feel for how a system should behave. In a very simple, lightly loaded system 29% might make sense; for more complex systems this seems like a lot.

A single debug bit provides a wealth of timing information. Another example is Figure 4-9, which shows an interrupt's latency. Though chip vendors spec interrupt latency in terms of the time the hardware needs to recognize the external event, to firmware folks a more useful measure is time-from-input to the time we're doing something useful, which may be many dozens of clock cycles. The multiple levels of vectoring needed by the average processor, plus important housekeeping such as context pushing, are all ultimately overhead incurred before the code starts doing something useful.

Unhappily, this definition is rather slippery, as it depends on the behavior of the entire system. An ISR that leaves interrupts disabled increases latency for every other task. Latency on a complex system is virtually impossible to predict, so take some measurements on time-critical interrupts.

The figure's bottom trace is the assertion of an active low interrupt. The top trace shows a debug bit the ISR drives high. Here we see almost 50 μsec of latency between the device requesting service and the ISR starting (measured as the time from /INTR falling to the debug bit rising).

Fifty microseconds again violates my commonsense feel for how systems should operate. The number may be right . . . or it may indicate that some other task is hogging time.

Real Time Means Right Now! 81

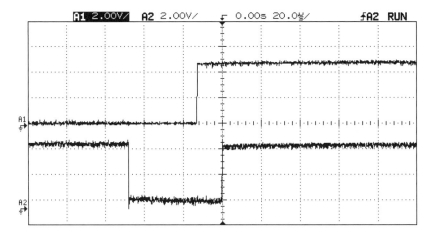

FIGURE 4-9 Measuring interrupt latency.

Perhaps an even more profound measurement is the system's total idle time. Is the CPU 100% loaded? 90%? Without this knowledge you cannot reliably tell the boss, "Sure, we can add that feature."

Instead of driving the debug bit in ISRs, toggle it in the idle loop. Applications based on RTOSs often don't use idle loops, so create a low-priority idle task that runs when there's nothing to do.

The instrumented idle loop looks like this:

```
idle:
    drive debug bit high
    drive debug bit low
    look for something to do
    jump to idle
```

While the idle loop runs, the debug bit toggles up and down at a high rate of speed (see Figure 4-10). If you turn the scope's time base down (to more time per division), the toggling bit looks more like hash (Figure 4-11), with long down periods indicating that the code is no longer in the idle loop. In this example about a third of the processing time is unused.

If an interrupt occurs after setting the bit high, but before returning it to zero, then the "busy" interval will look like a one on the scope and not the zero indicated in Figure 4-11. "Idle" times are those where you see hash—the signal rapidly cycling up and down. "Busy" times are those where the signal is a steady one or zero.

Too many developers fall into the serendipity school of debugging. They feel that if the system works and meets external specifications, it's

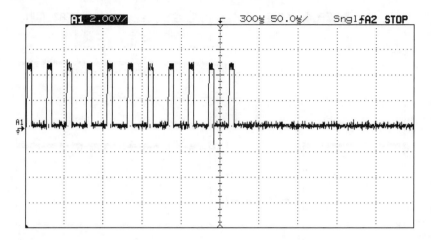

FIGURE 4-10 An idle loop quickly toggles the debug bit . . . until there's something to do!

ready to ship. Wrong. Hardware engineers stress their creations by running them over a temperature range. We should do the same, instrumenting our code or otherwise using performance-measuring tools, to be quite sure the system has sufficient margins. It's trivial to take quite accurate performance data.

The RTOS

Whenever an application manages multiple processes and devices, whenever one handles a variety of activities, an RTOS is a logical tool that lets us simplify the code and help it run better.

Consider the difficulty of building, say, a printer. Without an RTOS, one monolithic hunk of code would have to manage the door switches and paper feeding and communications and the print engine—all at the same time. Add an RTOS, and individual tasks each manage one of these activities; except for some status information, no task needs to know much about what any other one is doing. In this case the RTOS allows us to partition our code in the time domain (each of these activities is running concurrently) and procedurally (each task handles one thing).

An important truism of software engineering is that code complexity—and thus development time—grows much faster than program size. Any mechanism that segments the code into many small independent pieces reduces the complexity; after all, this is why we write with lots of functions and not one huge main() program. Clever partitioning yields bet-

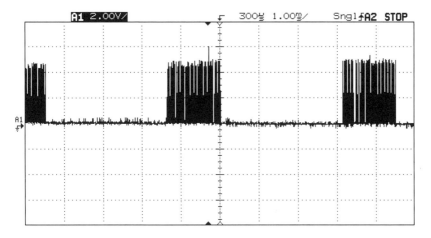

FIGURE 4-11 Measuring system idle time.

ter programs faster, and the RTOS is probably the most important way to partition code in the time dimension.

At its simplest level, an RTOS is a context switcher. You break your application into multiple tasks and allow the RTOS to execute the tasks in a manner determined by its scheduling algorithm. A round-robin scheduler typically allocates more or less fixed chunks of time to the tasks, executing each one for a few milliseconds or so before suspending it and going to the next ready task in the queue. In this way all tasks get their fair shot at some CPU time.

Another sort of scheduler is one using RMA—rate monotonic analysis. If the CPU is not completely performance bound, it's sometimes possible to guarantee hard real-time response by giving each task a priority inversely proportional to the task's period.

Regardless of scheduling mechanism, all RTOSs include priority schemes so you can statically and dynamically cause the context switcher to allocate more or less time to tasks. Important or time-critical activities get first shot at running. Less important housekeeping tasks run only as time allows. Your code sets the priorities; the RTOS takes care of starting and running the tasks.

If context switching were the only benefit of an RTOS, then none would be more than a few hundred bytes in size. Novice users all too often miss the importance of the sophisticated messaging mechanisms that are a standard part of all commercial operating systems. Queues and mailboxes let tasks communicate safely.

"Safely" is important, as global variables, the old standby of the desperate programmer, are generally a Bad Idea and are deadly in any interrupt-driven system. We all know how globals promote bugs by being available to every function in the code; with multitasking systems they lead to worse conflicts as several tasks may attempt to modify a global all at the same time.

Instead, the operating system's communications resources let you cleanly pass a message without fear of its corruption by other tasks. Properly implemented code lets you generate the real-time analogy of OOP's first tenet: encapsulation. Keep all of the task's data local, bound to the code itself, and hidden from the rest of the system.

For instance, one challenge faced by many embedded systems is managing system status info. Generally, lots and lots of different inputs, from door switches to the results of operator commands, affect total status. Maintain the status in a global data structure and you'll surely find it hammered by multiple tasks. Instead, bind the data to a task, and let other tasks set and query it via requests sent through queues or mailboxes.

Is this slower than using a global? Sure. It uses more memory, too. Just as we make some compromises in selecting a compiler over an assembler, proper use of an RTOS trades off a bit of raw CPU horsepower for better code that's easier to understand and maintain.

Most operating systems give you tools to manage resources. Surely it's a bad idea for multiple tasks to communicate with a UART or similar device simultaneously. One way to control this is to lock the resource—often using a semaphore or other RTOS-supplied mechanism—so only one task at a time can access the device.

Resource locking and priority systems lead to one of the perils of real-time systems: priority inversion. This is the deadly condition where a low-priority task blocks a ready and willing high-priority task.

Suppose the system is more or less idle. A background, perhaps unimportant, task asks for and gets exclusive access to a comm port. It's locked now, dedicated to the task until released. Suddenly an oh-my-god interrupt occurs that starts off the system's highest priority and most critical task. It, too, asks for exclusive comm port access, only to be denied that by the OS since the resource is already in use. The high-priority task is in control; the lower one can't run, and can't complete its activity and thus release the comm port. The least important activity of all has blocked the most important!

Most operating systems recognize the problem and provide a workaround. For example in VxWorks you can use their mutual exclusion semaphores to enable "priority inheritance." The task that locks the resource runs at the priority of the highest priority task that is blocked on the same resource. This permits the normally less important task to complete, so it can unlock the resource and allow the high-priority task to do its thing.

If you're not using an RTOS in your embedded designs today, you surely will be tomorrow. Get familiar with the concepts, as designing tasking code requires a somewhat different view—the time domain view—than conventional procedural programming. Check out Jean LaBrosse's free uC/OS; the companion book is as good an introduction to using an RTOS as you're likely to find. See www.ucos-ii.com.

Improvements to these tools come almost daily. Keep on top of the field to avoid the fate of the dinosaurs.

CHAPTER **5**

Firmware Musings

Hacking Peripheral Drivers

Experienced software engineers find no four-letter word more offensive than "hack." We believe that only amateurs, with more enthusiasm than skill, hack code.

Yet hacking is indeed a useful tool in limited circumstances.

This is not a rant against software methodologies—far from it. I think, though, a clever designer will identify risk areas and take steps to mitigate those risks *early* in a development program. Sometimes cranking code, maybe even lousy code, and diddling with it is the only way to figure out how to efficiently move forward.

No part of the firmware is more fraught with risks and unknowns than the peripheral drivers. *Don't* assume you are smart enough to create complex hardware drivers correctly the first time! Plan for problems instead of switching on the usual panic mode at debug time.

Before writing code, before playing with the hardware, build a shell of an executable using the tools allocated for the project. Use the same compiler, locator (if any), linker, and startup code. Create the simplest of programs, nothing more than the startup code and a null loop in main() (or its equivalent, when you're working in another language).

If the processor has internal chip-selects, figure out how to program these and include the setups in your startup code. Then, make the null loop work. This gives you confidence in the system's skeleton, and more importantly creates a backbone to plug test code into.

Next, create a single, operating, interrupt service routine. You're going to have to do this sooner or later anyway; swallow the bitter pill up front.

Identify every hardware device that needs a driver. This may even include memory, where (as with Flash) your code must do *something* to make it operate. Make a list, check it twice—LEDs, displays, timers, serial channels, DMA, communications controllers—include each component.

Surely you'll use a driver for each, though in some cases the driver may be segmented into several hunks of code, such as a couple of ISRs, a queue handler, and the like.

Next, set up a test environment for fiddling with the hardware. Use an emulator, a ROM monitor, or any tool that lets you start and stop the code. Manually exercise the ports (issue inputs and outputs to the device).

Gain mastery of each component by making it *do* something. Don't write code at this point—use your tool's input/output commands. If the port is a stack of LEDs, figure out how to toggle each one on and off. It's kind of fun, actually, to watch your machinations affect the hardware!

This is the time to develop a deep understanding of the device. All too often the documentation will be incomplete or just plain wrong. Bits inverted and transposed. Incorrect register addresses. You'll never find these problems via the normal design–code–inspect–debug cycle. Only playing with the devices—hacking!—with a decent debugging tool will unveil the peripheral's mysteries.

If you can't speak the hardware lingo, working with a part that has 100 "easy-to-set-up" registers will be impossible. If you are a hardware expert, dealing with these complex parts is merely a nightmare. Count on agony when the databook for a lousy timer weighs a couple of pounds.

Adopt a philosophy of creating a stimulus, then measuring the system's response with an appropriate tool.

Figures 5-1 and 5-2 illustrate this principle. The debugger's (in this case, driving an emulator) low-level commands configure the timer inside a 386EX. The response, measured on a scope, shows how the timer behaves with the indicated setup.

Using a serial port? Connect a terminal and learn how to transmit a single character. Again, manually set up the registers (carefully documenting what you did), using parameters extracted from the databook, using the tool's output command to send characters. Lots of things can go wrong with something as complicated as a UART, so I like to instrument its output with a scope. If the baud rate is incorrect, a terminal will merely display scrambled garbage; the scope will clearly show the problem.

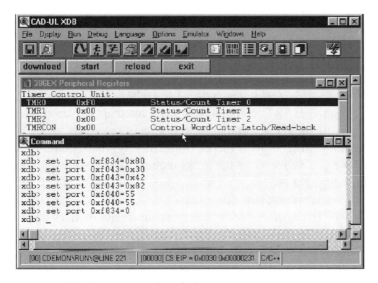

FIGURE 5-1 Hacking a peripheral driver.

Then write a shell of a driver in the selected language. Take the information gleaned from the databook and proven in your experiments to work, and codify it in code once and for all. Test the driver. Get it right!

Now you've successfully created a module that handles that hardware device.

Master one portion of a device at a time. On a UART, for example, figure out how to transmit characters reliably and document what you

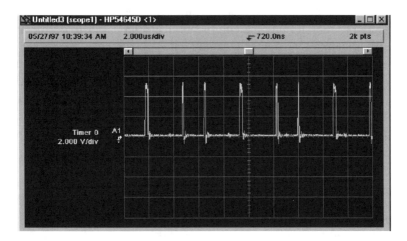

FIGURE 5-2 Hacking a peripheral driver.

did, before you move on to receiving. Segment the problem to keep things simple.

If only we could live with simple programmed inputs and outputs! Most nontrivial peripherals will operate in an interrupt-driven mode. Add ISRs, one at a time, testing each one, for each part of the device. For example, with the UART, completely master interrupt-driven transmission before moving on to interrupting reception.

Again, with each small success immediately create, compile, and test code before you've forgotten the tricks required to make the little beast operate properly. Databooks are cornucopias of information and misinformation; it's astonishing how often you'll find a bit documented incorrectly. Don't rely on frail memory to preserve this information. Mark up the book, create and test the code, and move on.

Some devices are simply too complex to yield to manual testing. An Ethernet driver or an IEEE-488 port both require so much setup that there's no choice but to initially write a lot of code to preset each internal register. These are the most frustrating sorts of devices to handle, as all too often there's little diagnostic feedback—you set a zillion registers, burn some incense, and hope it flies.

If your driver will transfer data using DMA, it still makes sense to first figure out how to use it a byte at a time in a programmed I/O mode. Be lazy—it's just too hard to master the DMA, interrupt completion routines, and the part itself all at once. Get single-byte transfers working before opening the Pandora's box of DMA.

In the "make it work" phase we usually succumb to temptation and hack away at the code, changing bits just to see what happens. The documentation generally suffers. Leave a bit of time before wrapping up each completed routine to tune the comments. It's a lot easier to do this when you still remember what happened and why.

More than once I've found that the code developed this way is ugly. Downright lousy, in fact, as coding discipline flew out the window during the bit-tweaking frenzy. The entire point of this effort is to master the device (first) and create a driver (second). Be willing to toss the code and build a less offensive second iteration. Test that too, before moving on.

Selecting Stack Size

With experience, one learns the standard, scientific way to compute the proper size for a stack: Pick a size at random and hope.

Unhappily, if your guess is too small the system will erratically and maybe infrequently crash in horrible ways. And RAM is still an expensive resource, so erring on the side of safety drives recurring costs up.

With an RTOS the problem is multiplied, since every task has its own stack.

It's feasible, though tedious, to compute stack requirements when coding in assembly language by counting calls and pushes. C—and even worse, C++—obscures these details. Runtime calls further distance our understanding of stack use. Recursion, of course, can blow stack requirements sky-high.

Any of a number of problems can cause the stack to grow to the point where the entire system crashes. It's tough to go back and analyze the failure after the crash, as the program will often write all over itself or the variables, removing all clues.

The best defense is a strong offense. Odds are your stack estimate will be wrong, so instrument the code from the very beginning so you'll know, for sure, just how much stack is needed.

In the startup code or whenever you define a task, fill the task's stack with a unique signature such as 0x55AA (Figure 5-3). Then, probe the stacks occasionally using your debugger and see just how many of the assigned locations have been used (the 0x55AA will be gone).

Knowledge is power.

Also consider building a stack monitor into your code. A stack monitor is just a few lines of assembly language that compares the stack pointer

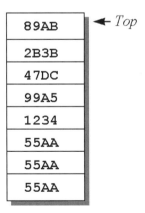

FIGURE 5-3 Proactively fill the stack with 0x55AA to find overrun problems. Note that the lower three words have been unused.

to some limit you've set. Estimate the total stack use, and then double or triple the size. Use this as the limit.

Put the stack monitor into one or more frequently called ISRs. Jump to a null routine, where a breakpoint is set, when the stack grows too big.

Be sure that the compare is "fuzzy." The stack pointer will never *exactly* match the limit.

By catching the problem *before* a complete crash, you can analyze the stack's contents to see what led up to the problem. You may see an ISR being interrupted constantly (that is, a lot of the stack's addresses belong to the ISR). This is a sure indication of code that's too slow to keep up with the interrupt rate. You can't simply leave interrupts disabled longer, as the system will start missing them. Optimize the algorithm and the code in that ISR.

The Curse of Malloc( )

Since the stack is a source of trouble, it's reasonable to be paranoid and not allocate buffers and other sizable data structures as automatics. Watch out! Malloc( ), a quite logical alternative, brings its own set of problems. A program that dynamically allocates and frees lots of memory—especially variably-sized blocks—will fragment the heap. At some point it's quite possible to have lots of free heap space, but so fragmented that malloc( ) fails.

If your code does not check the allocation routine's return code to detect this error, it will fail horribly. Of course, detecting the error will also no doubt result in a horrible failure, but gives you the opportunity to show an error code so you'll have a chance of understanding and fixing the problem.

If you chose to use malloc( ), *always* check the return value and safely crash (with diagnostic information) if it fails.

Garbage collection—which compacts the heap from time to time—is almost unknown in the embedded world. It's one of Java's strengths and weaknesses, as the time spent compacting the heap generally shuts down all tasking. Though there's lots of work going on developing real-time garbage collection, as of this writing there is no effective approach.

Sometimes an RTOS will provide alternative forms of malloc( ), which let you specify which of several heaps to use. If you can constrain your memory allocations to standard-sized blocks, and use one heap per size, fragmentation won't occur.

One option is to write a replacement function of the form pmalloc (heap_number). You defined a number of heaps, each one of which has a dedicated allocation size. Heap 1 might return a 2000-byte buffer, heap 2 100 bytes, and so on. You then constrain allocations to these standard-size blocks to eliminate the fragmentation problem.

When using C, if possible (depending on resource issues and processor limitations), always include Walter Bright's MEM package (www.snippets.org/mem.txt) with the code, at least for debugging. MEM provides the following:

- ISO/ANSI verification of allocation/reallocation functions
- Logging of all allocations and frees
- Verifications of frees
- Detection of pointer over- and under-runs
- Memory leak detection
- Pointer checking
- Out-of-memory handling

Banking

When asked how much money is enough, Nelson Rockefeller reportedly replied, "Just a little bit more." We poor folks may have trouble understanding his perspective, but all too often we exhibit the same response when picking the size of the address space for a new design. Given that the code inexorably grows to fill any allocated space, "just a little more" is a plea we hear from the software people all too often.

Is the solution to use 32-bit machines exclusively, cramming a full 4 GB of RAM into our cost-sensitive application in the hopes that no one could possibly use that much memory?

Though clearly most systems couldn't tolerate the costs associated with such a poor decision, an awful lot of designers take a middle tack, selecting high-end processors to cover their posterior parts.

A 32-bit CPU has tons of address space. A 16-bitter sports (generally) 1 to 16 Mb. It's hard to imagine needing more than 16 Mb for a typical embedded app; even 1 Mb is enough for the vast majority of designs.

A typical 8-bit processor, though, is limited to 64k. Once this was an ocean of memory we could never imagine filling. Now C compilers let us reasonably produce applications far more complex than we dreamed of even a few years ago. Today the midrange embedded systems I see usually burn up something between 64k and 256k of program and data space—too much for an 8-bitter to handle without some help.

If horsepower were not an issue, I'd simply toss in an 80188 and profit from the cheap 8-bit bus that runs 16-bit instructions over 1 Mb of address space. Sometimes this is simply not an option; an awful lot of us design upgrades to older systems. We're stuck with tens of thousands of lines of "legacy" code that are too expensive to change. The code forces us to continue using the same CPU. Like taxes, programs always get bigger, demanding more address space than the processor can handle.

Perhaps the only solution is to add address bits. Build an external mapper using PLDs or discrete logic. The mapper's outputs go into high-order address lines on your RAM and ROM devices. Add code to remap these lines, swapping sections of program or data in and out as required.

Logical to Physical

Add a mapper, though, and you'll suddenly be confronted with two distinct address spaces that complicate software design.

The first is the *physical* space—the entire universe of memory on your system. Expand your processor's 64k limit to 256k by adding two address lines, and the physical space is 256k.

Logical addresses are the ones generated by your program, and thence asserted onto the processor's bus. Executing a MOV A,(0FFFF) instruction tells the processor to read from the very last address in its 64k logical address space. External banking hardware can translate this to some other address, but the code itself remains blissfully unaware of such actions. All it knows is that some data comes from memory in response to the 0FFFF placed on the bus. The program can never generate a logical address larger than 64k (for a typical 8-bit CPU with 16 address lines).

This is very much like the situation faced by 80x86 assembly-language programmers: 64k segments are essentially logical spaces. You can't get to the rest of physical memory without doing *something*; in this case reloading a segment register.

Conversely, if there's no mapper, then the physical and logical spaces are identical.

Hardware Issues

Consider doubling your address space by taking advantage of processor cycle types. If the CPU differentiates memory reads from fetches, you may be able to easily produce separate data and code spaces. The 68000's seldom-used function codes are for just this purpose, potentially giving it distinct 16-Mb code and data spaces.

Writes should clearly go to the data area (you're not writing self-modifying code, are you?). Reads are more problematic. It's easy to distinguish memory reads from fetches when the processor generates a fetch signal for every instruction byte. Some processors (e.g., the Z80) produce a fetch only on the read of the first byte of a multiple byte opcode; subsequent ones all look the same as any data read. Forget trying to split the memory space if cycle types are not truly unique.

When such a space-splitting scheme is impossible, then build an external mapper that translates address lines. However, avoid the temptation to simply latch upper address lines. Though it's easy to store A16, A17, et al. in an output port, every time the latch changes the *entire* program gets mapped out. Though there are awkward ways to write code to deal with this, add a bit more hardware to ease the software team's job.

Design a circuit that maps just portions of the logical space in and out. Look at software requirements first to see what hardware configuration makes sense.

Every program needs access to a data area that holds the stack and miscellaneous variables. The stack, for sure, must always be visible to the processor so calls and returns function. Some amount of "common" program storage should always be mapped in. The remapping code, at least, should be stored here so that it doesn't disappear during a bank switch. Design the hardware so these regions are always available.

Is the address space limitation due to an excess of code or of data? Perhaps the code is tiny, but a gigantic array requires tons of RAM. Clearly, you'll be mapping RAM in and out, leaving one area of ROM— enough to store the entire program—always in view. An obese program yields just the opposite design. In either of these cases a logical address space split into three sections makes the most sense: common code (always visible, containing runtime routines called by a compiler and the mapping code), mapped code or data, and common RAM (stack and other critical variables needed all the time).

For example, perhaps 0000 to 03FFF is common code. 4000 to 7FFF might be banked code; depending on the setting of a port it could map to almost any physical address. 8000 to FFFF is then common RAM.

Sure, you can use heroic programming to simplify the hardware. I think it's a mistake, as the incremental parts cost is minuscule compared to the increased bug rate implicit in any complicated bit of code. It *is* possible—and reasonable—to remove one bank by copying the common code to RAM and executing it there, using one bank for both common code and data.

It's easy to implement a three-bank design. Suppose addresses are arranged as in the previous example. A0 to A14 go to the RAM, which is selected when A15 = 1.

Turn ROM on when A15 is low. Run A0 to A14 into the ROM. Assuming we're mapping a 128k × 8 ROM into the 32k logical space, generate a fake A15 and A16 (simple bits latched into an output port) that go to the ROM's A15 and A16 inputs. However, feed these through AND gates. Enable the gates only when A15 = 0 (RAM off) and A14 = 1 (bank area enabled).

RAM is, of course, selected with logical addresses between 8000 and FFFF. Any address under 4000 disables the gates and enables the first 4000 locations in ROM. When A14 is a one, whatever values you've stuck into the fake A15 and A16 select a chunk of ROM 4000 bytes long.

The virtue of this design is its great simplicity and its conservation of ROM—there are no wasted chunks of memory, a common problem with other mapping schemes.

Occasionally a designer directly generates chip selects (instead of extra address lines) from the mapping output port. I think this is a mistake. It complicates the ROM select logic. Worse, sometimes it's awfully hard to make your debugging tools understand the translation from addresses to symbols. By translating *addresses* you can provide your debugger with a logical-to-physical translation cheat sheet.

The Software

In assembly language you control everything, so handling banked memory is not too difficult. The hardest part of designing remappable code is figuring out how to segment the banks. Casual calling of other routines is out, as you dare not call something not mapped in.

Some folks write a bank manager that tracks which routines are currently located in the logical space. All calls, then, go through the bank manager, which dynamically brings routines in and out as needed.

If you were foresighted enough to design your system around a real-time operating system (RTOS), then managing the mapper is much simpler. Assign one task per bank. Modify the context switcher to remap whenever a new task is spawned or reawakened.

Many tasks are quite small—much smaller than the size of the logical banked area. Use memory more efficiently by giving tasks two banking parameters: the bank number associated with the task, and a starting offset into the bank. If the context switcher both remaps and then starts the task at the given offset, you'll be able to pack multiple tasks per bank.

Some C compilers come with built-in banking support. Check with your vendor. Some will completely manage a multiple bank system, automatically remapping as needed to bring code in and out of the logical address space. Figure on making a few patches to the supplied remapping code to accommodate your unique hardware design.

In C or assembly, using an RTOS or not, be sure to put all of your interrupt service routines and associated vectors in a common area. Put the banking code there as well, along with all frequently used functions (when you're using a compiler, put the entire runtime package in unmapped memory).

As always, when designing the hardware carefully document the approach you've selected. Include this information in the banking routine so some poor soul several years in the future has a fighting chance to figure out what you've done.

And, if you are using a banking scheme, be sure that the tools provide intelligent support. Quite a few 8-bit emulators, for example, do have extra address bits expressly for working in banked hardware. This means you can download code and even set breakpoints in banked areas that may not be currently mapped into the logical address space.

But be sure the emulator works properly with the compiler or assembler to give real source-level support in banked regions. If the compiler and emulator don't work together to share the physical and logical addresses of every line of code and every global/static variable, the "source" debugger will show nothing more useful than disassembled instructions. That's a terrible price to pay; in most cases you'll be well advised to find a more debuggable CPU.

Predicting ROM Requirements

It's rather astonishing how often we run into the same problem, yet take no action to deal with the issue once and for all. One common problem that drives managers wild is the old "running out of ROM space" routine—generally the week before shipping.

For two reasons it's very difficult to predict ROM requirements in the project's infancy. First, too many of us write code before we've done a complete and thoughtful analysis of the project's size. If you're not estimating code size (in lines of code or numbers of function points or a similar metric), then you're simply not a professional software engineer.

Second, we're generally not sure how to correlate a line of C to a number of bytes of machine code. Historical data is most useful if you've worked with the specific CPU and compiler in the past.

Regardless, when you start coding, maintain a spreadsheet that predicts the project's size. As a professional you've done the best possible job estimating the functions' sizes (in LOC, lines of code). List this data.

Whenever you complete a function, append the incremental size of the executable to the spreadsheet. Figure 5-4 shows an example, including each function, with estimated and actual LOC counts, and compiled sizes.

Any idiot—or at least any idiot with an engineering degree—can then write an equation that creates an average size of an LOC in bytes, and another that predicts total system size based on estimated LOC.

Make sure your calculations do not include the bare system skeleton—the C startup code and a null main() function—since the first line of C brings in the runtime package.

RAM Diagnostics

Beyond software errors lurks the specter of a hardware failure that causes our correct code to die, possibly creating a life-threatening horror, or maybe just infuriating a customer. Many of us write diagnostic code to help contain the problem. Much of the resulting code just does not address failure modes.

Obviously, a RAM problem will destroy most embedded systems. Errors reading from the stack will surely crash the code. Problems, especially intermittent ones, in the data areas may manifest bugs in subtle ways. Often you'd rather have a system that just doesn't boot, rather than one that occasionally returns incorrect answers.

Module	Est LOC	Act LOC	Size
Skeleton	300	310	21,123
RTOS		3423	11,872
TIMER_ISR	50	34	534
ATOD_ISR	75	58	798
TOD	120	114	998
PRINT_E	80	98	734
COMM_SER	90		
RD_ATOD	40		
	Bytes/LOC	4.01	
	Est Size	36580	

FIGURE 5-4 A spreadsheet that predicts ROM size.

Some embedded systems are pretty tolerant of memory problems. We hear of NASA spacecraft from time to time whose core or RAM develops a few bad bits, yet somehow the engineers patch their code to operate around the faulty areas, uploading the corrections over the distances of billions of miles.

Most of us work on systems with far less human intervention. There are no teams of highly trained personnel anxiously monitoring the health of each part of our products. It's our responsibility to build a system that works properly when the hardware is functional.

In some applications, though, a certain amount of self-diagnosis either makes sense or is required; critical life-support applications should use every diagnostic concept possible to avoid disaster due to a submicron RAM imperfection.

So, the first rule about diagnostics in general, and RAM tests in particular, is to clearly define your goals. Why run the test? What will the result be? Who will be the unlucky recipient of the bad news in the event an error is found, and what do you expect that person to do?

Will a RAM problem kill someone? If so, a very comprehensive test, run regularly, is mandatory.

Is such a failure merely a nuisance? For instance, if it keeps a cell phone from booting, if there's nothing the customer can do about the failure anyway, then perhaps there's no reason for doing a test. As a consumer I could care less why the damn phone stopped working . . . if it's dead, I'll take it in for repair or replacement.

Is production test—or even engineering test—the real motivation for writing diagnostic code? If so, then define exactly what problems you're looking for and write code that will find those sorts of troubles.

Next, inject a dose of reality into your evaluation. Remember that today's hardware is often very highly integrated. In the case of a microcontroller with on-board RAM, the chances of a memory failure that doesn't also kill the CPU is small. Again, if the system is a critical life-support application it may indeed make sense to run a test, as even a minuscule probability of a fault may spell disaster.

Does it make sense to ignore RAM failures? If your CPU has an illegal instruction trap, there's a pretty good chance that memory problems will cause a code crash you can capture and process. If the chip includes protection mechanisms (like the x86 protected mode), count on bad stack reads immediately causing protection faults your handlers can process. Perhaps RAM tests are simply not required, given these extra resources.

Inverting Bits

Most diagnostic code uses the simplest of tests—writing alternating 0x55 and 0xAA values to the entire memory array, and then reading the data to ensure that it remains accessible. It's a seductively easy approach that will find an occasional problem (like someone forgot to load all of the RAM chips), but that detects few real-world errors.

Remember that RAM is an array divided into columns and rows. Accesses require proper chip selects and addresses sent to the array—and not a lot more. The 0x55/0xAA symmetrical pattern repeats massively all over the array; accessing problems (often more common than defective bits in the chips themselves) will create references to incorrect locations, yet almost certainly will return what appears to be correct data.

Consider the physical implementation of memory in your embedded system. The processor drives address and data lines to RAM—in a 16-bit system there will surely be at least 32 of these. Any short or open on this huge bus will create bad RAM accesses. Problems with the PC board are far more common than internal chip defects, yet the 0x55/0xAA test is singularly poor at picking up these, the most likely, failures.

Yet the simplicity of this test and its very rapid execution have made it an old standby that's used much too often. Isn't there an equally simple approach that will pick up more problems?

If your goal is to detect the most common faults (PCB wiring errors and chip failures more substantial than a few bad bits here or there), then indeed there is. Create a short string of almost random bytes that you repeatedly send to the array until all of memory is written. Then, read the array and compare against the original string.

I use the phrase "almost random" facetiously, but in fact it hardly matters what the string is, as long as it contains a variety of values. It's best to include the pathological cases, such as 00, 0xaa, 0x55, and 0xff. The string is something you pick when writing the code, so it is truly not random, but other than these four specific values, you fill the rest of it with nearly any set of values, since we're just checking basic write/read functions (remember: memory tends to fail in fairly dramatic ways). I like to use very orthogonal values—those with lots of bits changing between successive string members—to create big noise spikes on the data lines.

To make sure this test picks up addressing problems, ensure that the string's length is not a factor of the length of the memory array. In other words, you don't want the string to be aligned on the same low-order addresses, which might cause an address error to go undetected. Since the string is much shorter than the length of the RAM array, you ensure that it repeats at a rate that is not related to the row/column configuration of the chips.

For 64k of RAM, a string 257 bytes long is perfect: 257 is prime, and its square is greater than the size of the RAM array. Each instance of the string will start on a different low-order address. Also, 257 has another special magic: you can include every byte value (00 to 0xff) in the string without effort. Instead of manually creating a string in your code, build it in real time by incrementing a counter that overflows at 8 bits.

Critical to this, and every other RAM test algorithm, is that you write the pattern to all of RAM before doing the read test. Some people like to do nondestructive RAM tests by testing one location at a time, then restoring that location's value, before moving on to the next one. Do this and you'll be unable to detect even the most trivial addressing problem.

This algorithm writes and reads every RAM location once, so it's quite fast. Improve the speed even more by skipping bytes, perhaps writing and reading every 3rd or 5th entry. The test will be a bit less robust, yet will still find most PCB and many RAM failures.

Some folks like to run a test that exercises each and every bit in their RAM array. Though I remain skeptical of the need, since most semiconductor RAM problems are rather catastrophic, if you do feel compelled to run such a test, consider adding another iteration of the algorithm just described, with all of the data bits inverted.

Noise Issues

Large RAM arrays are a constant source of reliability problems. It's indeed quite difficult to design the perfect RAM system, especially with the minimal margins and high speeds of today's 16- and 32-bit systems. If your system uses more than a couple of RAM parts, count on spending some time qualifying its reliability via the normal hardware diagnostic procedures. Create software RAM tests that hammer the array mercilessly.

Probably one of the most common forms of reliability problems with RAM arrays is pattern sensitivity. Now, this is not the famous pattern problems of yore, where the chips (particularly DRAMs) were sensitive to the groupings of ones and zeroes. Today the chips are just about perfect in this regard. No, today pattern problems come from poor electrical characteristics of the PC board, decoupling problems, electrical noise, and inadequate drive electronics.

PC boards were once nothing more than wiring platforms, slabs of tracks that propagated signals with near-perfect fidelity. With very high-speed signals, and edge rates (the time it takes a signal to go from a zero to a one or back) under a nanosecond, the PCB itself assumes all of the characteristics of an electronic component—one whose virtues are almost all problematic. It's a big subject [read *High Speed Digital Design—A Handbook of Black Magic*, by Howard Johnson and Martin Graham (1993 PTR Prentice Hall, NJ) for the canonical words of wisdom on this subject], but suffice it to say that a poorly designed PCB will create RAM reliability problems.

Equally important are the decoupling capacitors chosen, as well as their placement. Inadequate decoupling will create reliability problems as well.

Modern DRAM arrays are massively capacitive. Each address line might drive dozens of chips, with 5 to 10 pF of loading per chip. At high speeds the drive electronics must somehow drag all of these pseudo-capacitors up and down with little signal degradation. Not an easy job! Again, poorly designed drivers will make your system unreliable.

Electrical noise is another reliability culprit, sometimes in unexpected ways. For instance, CPUs with multiplexed address/data buses use external address latches to demux the bus. A signal, usually named ALE (Address Latch Enable) or AS (Address Strobe), drives the clock to these latches. The tiniest, most miserable amount of noise on ALE/AS will surely, at the time of maximum inconvenience, latch the data part of the cycle instead of the address. Other signals are also vulnerable to small noise spikes.

Unhappily, all too often common RAM tests show no problem when hidden demons are indeed lurking. The algorithm I've described, as well as most of the others commonly used, trade off speed against comprehensiveness. They don't pound on the hardware in a way designed to find noise and timing problems.

Digital systems are most susceptible to noise when large numbers of bits change all at once. This fact was exploited for data communications long ago with the invention of the Gray code, a variant of binary counting where no more than one bit changes between codes. Your worst nightmares of RAM reliability occur when all of the address and/or data bits change suddenly from zeroes to ones.

For the sake of engineering testing, write RAM test code that exploits this known vulnerability. Write 0xffff to 0x0000 and then to 0xffff, and do a read-back test. Then write zeroes. Repeat as fast as your loop will let you go.

Depending on your CPU, the worst locations might be at 0x00ff and 0x0100, especially on 8-bit processors that multiplex just the lower 8 address lines. Hit these combinations hard as well.

Other addresses often exhibit similar pathological behavior. Try 0x5555 and 0xaaaa, which also have complementary bit patterns.

The trick is to write these patterns back-to-back. Don't test all of RAM, with the understanding that both 0x0000 and 0xffff will show up in the test. You'll stress the system most effectively by driving the bus massively up and down all at once.

Don't even think about writing this sort of code in C. Any high-level language will inject too many instructions between those that move the bits up and down. Even in assembly the processor will have to do fetch cycles from wherever the code happens to be, which will slow down the pounding and make it a bit less effective.

There are some tricks, though. On a CPU with a prefetcher (all x86, 68k, etc.) try to fill the execution pipeline with code, so the processor does back-to-back writes or reads at the addresses you're trying to hit. And, use memory-to-memory transfers when possible. For example:

```
mov     si,0xaaaa
mov     di,0x5555
mov     [si],0xff
mov     [di],[si]           ; read ff00 from 0aaaa
                            ; and then write it
                            ; to 05555
```

DRAMs have memories rather like mine—after 2 to 4 milliseconds go by, they will probably forget unless external circuitry nudges them with a gentle reminder. This is known as "refreshing" the devices and is a critical part of every DRAM-based circuit extant.

More and more processors include built-in refresh generators, but plenty of others still rely on rather complex external circuitry. Any failure in the refresh system is a disaster.

Any RAM test should pick up a refresh fault—shouldn't it? After all, it will surely take a *lot* longer than 2–4 msec to write out all of the test values to even a 64k array.

Unfortunately, refresh is basically the process of cycling address lines to the DRAMs. A completely dead refresh system won't show up with the test indicated, since the processor will be merrily cycling address lines like crazy as it writes and reads the devices. There's no chance the test will find the problem. This is the worst possible situation: the process of running the test camouflages the failure!

The solution is simple: After writing to all of memory, just stop toggling those pesky address lines for a while. Run a tight do-nothing loop for a while (*very* tight . . . the more instructions you execute per iteration, the more address lines will toggle), and only then do the read test. Reads will fail if the refresh logic isn't doing its thing.

Though DRAMs are typically specified at a 2- to 4-msec maximum refresh interval, some hold their data for surprisingly long times. When memories were smaller and cells larger, each had so much capacitance that you could sometimes go for dozens of seconds without losing a bit. Today's smaller cells are less tolerant of refresh problems, so a 1- to 2-second delay is probably adequate.

A Few Notes on Software Prototyping

As a teenaged electronics technician I worked for a terribly undercapitalized small company that always spent tomorrow's money on today's problems. There was no spare cash to cover risks. As is so often the case, business issues overrode common sense and the laws of physics: all prototypes simply had to work, and were in fact shipped to customers.

Years ago I carried this same dysfunctional approach to my own business. We prototyped products, of course, but did so leaving no room for failure. Schedules had no slack; spare parts were scarce, and people heroically overcame resource problems. In retrospect this seems silly, since by definition we create prototypes simply because we expect mistakes, problems, and, well . . . failure.

Can you imagine being a civil engineer? Their creations—a bridge, a building, a major interchange—are all one-off designs that simply *must* work correctly the first time. We digital folks have the wonderful luxury of building and discarding trial systems.

Software, though, looks a lot like the civil engineer's bridge. Costs and time pressures mean that code prototypes are all too rare. We write the code and knock out most of the bugs. Version 1.0 is no more than a first draft, minus most of the problems.

Though many authors suggest developing version 1.0 of the software, then chucking it and doing it again, now correctly, based on what was learned from the first go-around, I doubt that many of us will often have that opportunity. The 1990s are just too frantic, workforces too thin, and time-to-market pressures too intense. The old engineering adage "If the damn thing works at all, ship it," once only a joke, now seems to be the industry's mantra.

Besides—who wants to redo a project? Most of us love the challenge of making something work, but want to move on to bigger and better things, not repeat our earlier efforts.

Even hardware is moving away from conventional prototypes. Reprogrammable logic means that the hardware is nothing more than software. Slap some smart chips on the board and build the first production run. You can (hopefully) tune the equations to make the system work despite interconnect problems.

We're paid to develop firmware that is correct—or at least correct enough—to form a final product, first time, every time. We're the high-tech civil engineers, though at least we have the luxury of fixing mistakes in our creations before releasing the product to the cruel world of users.

Though we're supposed to build the system right the first time, we're caught in a struggle between the computer's need for perfect instructions, and marketing's less-than-clear product definitions. The B-schools are woefully deficient in teaching their students—the future product definers—about the harsh realities of working in today's technological environment. Vague handwaving and whiteboard sketches are not a product spec. They need to understand that programmers must be unfailingly precise and complete in designing the code. Without a clear spec, the programmers themselves, by default, must create the spec.

Most of us have heard the "but *that's* not what I wanted" response from management when we demo our latest creation. All too often the customer—management, your boss, or the end user—doesn't really know what they want until they see a working system. It's clearly a Catch-22 situation.

The solution is a prototype of the system's software, running a minimal subset of the application's functionality. This is not a skeleton of the final code, waiting to be fleshed out after management puts in their two cents. I'm talking about truly disposable code.

Most embedded systems do possess some sort of look and feel, despite the absence of a GUI. Even the light-up sneakers kids wear (which, I'm told, use a microcontroller from Microchip) have at least a "look." How long should the light be on? Is it a function of acceleration? If I were designing such a product, I'd run a cable from the sneaker to a development system so I could change the LED's parameters in seconds while the MBAs argue over the correct settings.

"Wait," you say. "We can't do that here! We *always* ship our code!" Though this is the norm, I'm running into more and more embedded developers who have been so badly burned by inadequate/incorrect specifications that even management grudgingly backs up their rapid prototyping efforts. However, any prototype will fail unless the goals are clearly spelled out.

The best prototype spec is one that models risk factors in the final product. Risk comes in far too many flavors: user interface (human interaction with the unit, response speed), development problems (tools, code speed, code size, people skill sets), "science" issues (algorithms, data reduction, sampling intervals), final system cost (some complex sum of engineering and manufacturing costs), time to market, and probably other items as well.

A prototype may not be the appropriate vehicle for dealing with all risk factors. For example, without building the real system it'll be tough to extrapolate code speed and size from any prototype.

The first ground rule is to define the result you're looking for. Is it to perfect a data reduction algorithm? To get consensus on a user interface? Focus with unerring intensity on just that result. Ignore all side issues. Build just enough code to get the desired result. Real systems need a spec that defines what the product does; a rapid prototype needs a spec that spells out what *won't* be in it.

More than anything you need a boss who shields you from creeping featurism. We know that a changing spec is the bane of real systems; surely it's even more of a problem in a quick-turn model system.

Then you'll need an understanding of what decisions will be made as a result of the prototype. If the user interface will be pretty much constant no matter what turns up in the modeling phase, hey—just jump into final product development. If you know the answer, don't ask the question!

Define the deadline. Get a prototype up and running at warp speed. Six months or a year of fiddling around on a model is simply too long. The raison d'être for the prototype is to identify problems and make changes. Get these decisions made early by producing something in days or weeks. Develop a schedule with many milestones where nondevelopers get a chance to look at the product and fiddle with it a bit.

For a prototype where speed and code size are not a problem, I like to use really high-level "languages" like Basic. Excel. Word macros. The goal is to get something going *now*. Use every tool, no matter how much it offends your sensibilities, to accomplish that mission.

Does your product have a GUI? Maybe a control panel? Look at products like those available from National Instruments and IoTech. These companies provide software that lets you produce "virtual instruments" by clicking and dragging knobs, displays, and switches around on a PC's screen. Couple that to standard data acquisition boards and a bit of code in Basic or C, and you can produce models of many sorts of embedded systems in hours.

The cost of creating a virtual model of your product, using purchased components, is immeasurably small compared to that of designing, building, and troubleshooting real hardware and software. Though there's no way to avoid building hardware at some point, count on adding months to a project when a new board design is required.

Another nice feature of doing a virtual model of the product is the certainty of creating worthless code. You'll focus on the real issues—the ones identified in your prototyping goals—and not the problems of creating documented, portable, well-structured software. The code will be no more than the means to the end. You'll toss the code as casually as the hardware folks toss prototype PC boards.

I mentioned using Excel. Spreadsheets are wonderful tools for evaluating the product's science. Unsure about the behavior of a data-smoothing algorithm? Fiddling with a fuzzy-logic design? Wondering how much precision to carry? Create a data set and put it in your trusty spreadsheet. Change the math in seconds; graph the results to see what happens. Too many developers write a ton of embedded code, only to spend months tuning algorithms in the unforgiving environment of an 8051 with limited memory.

Though a spreadsheet masks the calculations' speed, you can indeed get some sort of final complexity estimate by examining the equations. If the algorithm looks terribly slow, work within the forgiving environment of the spreadsheet to develop a faster approach. We all know, though too often ignore, the truth that the best performance enhancements come from tuning the algorithm, not the code.

Though the PC is a great platform for modeling, do consider using current company products as prototype platforms. Often new products are derivatives of older ones. You may have a lot of extant hardware and software—that works!—in a system on the shelf. Be creative and use every resource available to get the prototype up and running.

Toss out the standards manual. Use every trick in the book to get it done *fast*. Do code in small functions to get something testable quickly, and to minimize the possibility of making big mistakes.

There's a secret benefit to using cruddy "languages" for software prototypes: write your proto code in Visual Basic, say, and no matter how hard management screams, it simply cannot be whisked off into the product as final code. Clever language selection can break the dysfunctional last-minute conversion of test code to final firmware.

All of us have worked with that creative genius who can build anything, who pounds out a thousand lines of code a day, but who can never seem to complete a project. Worse—the fast coder who spends eons debugging the megabyte of firmware he wrote on a Jolt-driven all-nighter. Then there are the folks who produce working code devoid of documentation, who develop rashes or turn into Mr. Hyde when told to add comments.

We struggle with these folks, plead with them, send them to seminars, lead by example, all too often without success. Some of them are prima donnas who should probably get the ax. Others are really quite good, but simply lack the ability to deal with detail ... which is essential since, in a released product, every lousy bit must be right.

These are the ideal prototype developers. Bugs aren't a big issue in a model, and documentation is less than important. The prototype lets them exercise their creative zeal, while its limited scope means that problems are not important. Toss Twinkies and caffeine into their lair and stand back. You'll get your system fast, and they'll be happy employees. Use the more disciplined team members to get the bugless real product to market.

Part of management is effectively using people's strengths while mitigating their weaknesses. Part of it is also giving the workers a break once in a while. No one can crank out 70-hour weeks forever without cracking.

CHAPTER **6**

Hardware Musings

Debuggable Designs

An unhappy reality of our business is that we'll surely spend lots of time—far too much time—debugging both hardware and firmware. For better or worse, debugging consumes project-months with reckless abandon. It's usually a prime cause of schedule collapse, disgruntled team members, and excess stomach acid.

Yet debugging will never go away. Practicing even the very best design techniques will never eliminate mistakes. No one is smart enough to anticipate every nuance and implication of each design decision on even a simple little 4k 8051 product; when complexity soars to hundreds of thousands of lines of code coupled to complex custom ASICs we can only be sure that bugs will multiply like rabbits.

We know, then, up front when making basic design decisions that in weeks or months our grand scheme will go from paper scribbles to hardware and software ready for testing. It behooves us to be quite careful with those initial choices we make, to be sure that the resulting design isn't an undebuggable mess.

Test Points Galore

Always remember that, whether you're working on hardware or firmware problems, the oscilloscope is one of the most useful of all debugging tools. A scope gives instant insight into difficult code issues such as operation of I/O ports, ISR sequencing, and performance problems.

109

Yet it's tough to probe modern surface-mount designs. Those tiny whisker-thin pins are hard enough to see, let alone probe. Drink a bit of coffee and you'll dither the scope connection across three or four pins.

The most difficult connection problem of all is getting a good ground. With speeds rocketing toward infinity the scope will show garbage without a short, well-connected ground, yet this is almost impossible when the IC's pin is finer than a spiderweb.

So, when laying out the PCB add lots of ground points scattered all over the board. You might configure these to accept a formal test point. Or, simply put holes on the board, holes connected to the ground plane and sized to accept a resistor lead. Before starting your tests, solder resistors into each hole and cut off the resistor itself, leaving just a half-inch stub of stiff wire protruding from the board. Hook the scope's oversized ground clip lead to the nearest convenient stub.

Figure on adding test points for the firmware as well. For example, the easiest way to measure the execution time of a short routine is to toggle a bit up for the duration of the function. If possible, add a couple of parallel I/O bits just in case you need to instrument the code.

Add test points for the critical signals you know will be a problem. For example:

- Boot loads are always a problem with downloadable devices (Flash, ROM-loaded FPGAs, etc.). Put test points on the critical load signals, as you'll surely wrestle with these a bit.
- The basic system timing signals all need test points: read, write, maybe wait, clock, and perhaps CPU status outputs. All system timing is referenced to these, so you'll surely leave probes connected to those signals for days on end.
- Using a watchdog timer? Always put a test point on the time-out signal. Better, use an LED on a latch. You've *got* to know when the watchdog goes off, as this indicates a serious problem. Similarly, add a jumper to disable the watchdog, as you'll surely want it off when working on the code.
- With complex power-management strategies, it's a good idea to put test points on the reset pin, battery signals, and the like.

When using PLDs and FPGAs, remember that these devices incorporate all of the evils of embedded systems with none of the remedies we normally use: the entire design, perhaps consisting of tens of thousands of gates, is buried behind a few tens of pins. There's no good way to get "inside the box" and see what happens.

Some of these devices do support a bit of limited debugging using a serial connection to a pseudo-debug port. In such a case, by all means add the standard connector to your PCB! Your design will not work right off the bat; take advantage of any opportunity to get visibility into the part.

Also plan to dedicate a pin or two in each FPGA/PLD for debugging. Bring the pins to test points. You can always change the logic inside the part to route critical signal to these test points, giving you some limited ability to view the device's operation.

Similarly, if the CPU has a BDM or JTAG debugging interface, put a BDM/JTAG connector on the PCB, even if you're using the very best emulators. For almost zero cost you may save the project when/if the ICE gives trouble.

Very small systems often just don't have room for a handful of test points. The cost of extra holes on ultra-cheap products might be prohibitive. I always like to figure on building a real, honest, prototype first, one that might be a bit bigger and more expensive than the production version. The cost of doing an extra PCB revision (typically $1000 to $2000 for 5-day turnaround) is vanishingly small compared to your salary!

When management screams about the cost of test points and extra connectors, remember that you do not have to load these components during the production run. Install them on the prototypes, leaving them off the bill of materials. Years later, when the production folks wonder about all of the extra holes, you can knowingly smile and remember how they once saved your butt.

Resistors

When I was a young technician, my associates and I arrogantly believed we could build anything with enough 10k resistors and duct tape. Now it seems that even simple electronic toys use several million transistors encased in tiny SMT packages with hundreds of hairlike leads; no one talks about discrete components anymore. Yet no matter how digital our embedded designs get, we can never avoid certain fundamental electrical properties of our circuits.

For example, somehow the digital age has an ever-increasing need for resistors—so many, in fact, that most "discrete" resistors are now usually implemented in a monolithic structure, like an SIP, not so different from the ICs they are tied to.

Too often we spend our time carefully analyzing the best way to use a modern miracle of integration only to casually select discrete components because they are, well, *boring*. Who can get worked up over the lowly carbon resistor? You can't even buy them one at a time any more. At Radio Shack they come paired in bright decorator packages for an outrageous sum.

Back when I was in the emulator business we dealt with a lot of user target systems that, because of poor resistor choices, drove the tools out of their minds. Consider one typical example: a unit based on an 8-MHz 80188, memory and I/O all connected in a carefully thought-out manner. Power and ground distribution were well planned; noise levels were satisfyingly low. And yet . . . the only tool that seemed to work for debugging code was a logic analyzer. Every emulator the poor designer tested failed to run the code properly. Even a ROM emulator gave erratic results.

Though the emulator wouldn't run the user's code, it did show an immediate service of the non-maskable interrupt—which wasn't used in the system. (*Note: When things get weird, always turn to your emulator's trace feature, which will capture weirdness like no other tool.*)

A little further investigation revealed that the NMI input (which is active high on the 188) was tied low through a 47k resistor.

Now, the system ran fine with a ROM and processor on the board. I suppose the 47k pull-down was at least technically legitimate. A few microamps of leakage current out of the input pin through 47k yields a nice legal logic zero. Yet this 47k was too much resistance when any sort of tool was installed, because of the inevitable increase in leakage current.

Was the design correct because it violated none of Intel's design specs? I maintain that the specs are just the starting point of good design practice. Never, ever, violate one. Never, ever, assume that simply meeting spec is adequate.

A design is *correct* only if it reliably satisfies all intended applications—including the first of all applications, debugging hardware and software. If something that is technically correct prevents proper debugging, then there is surely a problem.

Pull-down resistors are often a source of trouble. It's practically impossible to pull down an LS input (leakage is so high the resistor value must be frighteningly low). Though CMOS inputs leak very little, you must be aware of every potential application of the circuit, including that of plugging tools in. The solution is to avoid pull-downs wherever possible.

In the case of a critical edge-triggered (read "really noise sensitive") input such as NMI, you simply should never pull it low. Tie it to ground. Otherwise, switching noise may get coupled into the input. Even worse, every time you lay out the PC board, the magnitude of the noise problem can change as the tracks move around the board.

Be conservative in your designs, especially when a conservative approach has no downside. If any input must be zero all of the time, simply tie it to ground and never again worry about it. I think folks are so used to adding pull-ups all over their boards that they design in pull-downs through the force of habit.

Once in a while the logic may indeed need a pull-down to deal with unusual I/O bits. Try to come up with a better design.

(The only exception is when you plan to use automatic test equipment to diagnose board faults. ATE gear injects signals into each node, so you'll often need to use a resistor pull-down in place of a ground. Use a small—really small, like 220 ohms—value.)

Though pull-downs are always problematic, well-designed boards use plenty of pull-up resistors—some to bias unused inputs, others to deal with signals and busses that tristate, and some to put switches and other inputs into known one states.

The biggest problem with pull-ups is using values that are too low. A 100k pull-up will in fact bias that CMOS gate properly, but creates a circuit with a terribly high impedance. Why not change to 10k? You buy an order of magnitude improvement in impedance and noise immunity, yet typically use no additional current since the gate requires only microamps of bias.

Vcc from a decent power supply is essentially a low-impedance connection to ground. Connect a 100k pull-up to a CMOS gate and the input is 100k away from ground, power, and everything else—you can overcome a 100k resistance by touching the net with a finger. A 10k resistor will overpower any sort of leakage created by fingers, humidity, and other effects.

Besides, that low-impedance connection will maintain a proper state no matter what tools you use. In the case of NMI from the example above, the tools weakly pulled NMI high so they could run standalone (without the target); the 47k resistor was too high a value to overcome this slight amount of bias.

If you are pulling up a signal from off-board, by all means use a very low value of resistance. The pull-up can act as a termination as well as a provider of a logic one, but the characteristic impedance of any cable is usually on the order of hundreds of ohms. A 100k pull-up is just too high to provide any sort of termination, leaving the input subject to cross coupling and noise from other sources. A 1k resistor will help eliminate transients and crosstalk.

Remember that you may not have a good idea what the capacitance of the wiring and other connections will be. A strong pull-up will reduce capacitive time constant effects.

Unused Inputs

Once upon a time, back before CMOS logic was so prevalent, you could often leave unused inputs dangling unconnected and reasonably expect to get a logic one. Still, engineers are a conservative lot, and most were careful to tie these spare pins to logic one or zero conditions.

But what exactly is a logic one? With 74LS logic it's unwise to use Vcc as an input to any gate. Most LS devices will happily tolerate up to 7 volts on Vcc before something fails, while the input pins have an absolute maximum rating of around 5.5 volts. Connecting an input to Vcc creates a circuit where small power glitches that the devices can tolerate may blow input transistors. It's far better (when using LS) to connect the input to Vcc through a resistor, thus limiting input current and yielding a more power-tolerant design.

Modern CMOS logic in most of its guises has the same absolute maximum rating for Vcc as for the inputs, so it's perfectly reasonable to connect input pins directly to Vcc—if you're sure that production will never substitute an LS equivalent for the device you've called out.

CMOS does require that every unused input be pulled to a valid logic zero or one to avoid generating an SCR latchup condition.

Fast CMOS logic (like 74FCT) switches so quickly, even at very low clock rates, that glitches with Fourier components into billions of cycles per second are not uncommon. Reduce noise susceptibility by tying your logic zeroes and ones directly to the power and ground planes.

And yet . . . one must balance the rules of good design with practical ways to make a debuggable system. A thousand years ago circuits used vacuum tubes mounted on a metal chassis. All connections were made by point-to-point wiring, so making engineering changes during prototype checkout must have been pretty easy. Later, transistors and ICs lived on PC boards, but incorporating modifications was still pretty simple. Now we're faced with whisker-thin leads on surface-mount components, with 8- and 10-layer boards where most tracks are buried under layers of epoxy and out of reach of our X-Acto knives. If we tie every unused input, even on our spare gates, to a solid power or ground connection, it'll be awfully hard to cut the connection free to tie it somewhere else. Lifting the pins on those spare gates might be a nightmare.

One solution is to build the prototype boards a little differently than the production versions. I look at a design and try to identify areas most likely to require cutting and pasting during checkout. A prime example is the programmable device—PALs or FPGAs or whatever. Bitter experience has taught me that probably I'll forget a crucial input to that PAL, or that I'll need to generate some nastily complex waveform using a spare output on the FPGA.

Some engineers figure that if they socket the programmable logic, they can lift pins and tack wires to the dangling input or output. I hate this solution. Sometimes it takes an embarrassing number of tries to get a complex PAL right—each time you must remove the device, bend the leads back to program it, and then reinstall the mods. (An alternative is to put a socket in the socket and lift the upper socket's leads.) When the device is PLCC or another, non-DIP package, it's even harder to get access to the pins.

So I leave all unused inputs on these devices unconnected when building the prototype, unfortunately creating a window of vulnerability to SCR latchup conditions. Then it's easy to connect mod wires to the unconnected pins. When the first prototype is done I'll change the schematic to properly tie off the unused inputs so prototype 2 (or the production unit) is designed correctly.

In years of doing this I have never suffered a problem from SCR latchup due to these dangling pins. The risk is always there, lurking and waiting for an unusual ESD or perhaps even a careless ungrounded finger biasing an input.

I do tie spare gate inputs to ground, even with the first run of boards. It just feels a little too dangerous to leave an unconnected 74HC74 lead dangling. However, if at all possible, I have the person doing the PCB layout connect these grounds on the bottom layer so that a few quick strokes of the X-Acto knife can free them to solve another "whoops."

In designs that use through-hole parts, by all means leave just a little extra room around each chip so you can socket the parts on the prototype. It's a lot easier to pull a connected pin from a socket than to cut it free from the board.

Clocks

For a number of years embedded systems lived in a wonderful era of compatibility. Just about all the signals on any logic board were relatively slow and generally TTL compatible. This lulled designers into a feeling of security, until far too many of us started throwing digital ICs together without considering their electrical characteristics. If a one is 2.4 volts and a zero 0.7, if we obey simple fanout rules, and as long as speeds are under 10 MHz or so, this casual design philosophy works pretty well. Unfortunately, today's systems are not so benign.

In fact, few microprocessors have *ever* exclusively used TTL levels. Surprise! Pull out a data sheet on virtually any microprocessor and look at the electrical specs page—you know, the section without coffee spills or solder stains. Skip over those 300 tattered pages about programming internal peripherals, bypass the pizza-smeared pinout section, and really look at those one or two pristine pages of DC specifications.

Most CPUs accept TTL-level data and control inputs. Few are happy with TTL on the clock and/or reset inputs. Each chip has different requirements, but in a quick look through the data books I came up with the following:

- 8086: Minimum Vih on clock: Vcc − 0.8
- 386: Minimum Vih on clock: Vcc − 0.8 at 20 MHz, 3.7 volts at 25 and 33 MHz
- Z80: Minimum Vih on clock: Vcc − 0.6
- 8051: Minimum Vih on clock and reset: 2.5 volts

In other words, connect your clock and maybe reset input to a normal TTL driver, and the CPU is out of spec. The really bad news is that these chips are manufactured to behave far better than the specs, so often they'll run fine despite illegal inputs. If only they failed immediately on any violation of specifications! Then, we'd find these elusive problems in the lab, long before shipping a thousand units into the field.

Fully 75% of the systems I see that use a clock oscillator (rather than a crystal) violate the clock minimum high-voltage requirement. It's scary to think we're building a civilization around embedded systems that, well, may be largely misdesigned.

If you drive your processor's clock with the output of a gate or flip-flop, be sure to use a device with true CMOS voltage levels. 74HCT or 74ACT/FCT are good choices. Don't even consider using 74LS without at least a heavy-duty pull-up resistor.

Those little 14-pin silver cans containing a complete oscillator are a good choice . . . if you read the data sheet first. Many provide TTL levels only. I'm not trying to be alarmist here, but look in the latest DigiKey catalog—they sell dozens of varieties of CMOS *and* TTL parts.

Clocks must be clean. Noise will cause all sorts of grief on this most important signal. It's natural to want to use a Thevenin termination to more or less match impedance on a clock routed over a long PCB trace or even off board. Beware! Thevenin terminations (typically a 220-ohm resistor to +5 and a 270 to ground) will convert your carefully crafted CMOS level to TTL.

Use series damping resistors to reduce the edge rate if noise is a problem. A pull-up might help with impedance matching if the power supply has a low impedance (as it should).

A better solution is to use clock-shaping logic near the processor itself. If the clock is generated a long way away, use a CMOS hysteresis circuit (such as a 74HCT14) to clean it up. The extra logic adds delay, though. If your system requires clock synchronization, then use a special low-skew clock driver made for that purpose.

In slower systems—under 20 MHz or so—I prefer to design circuits that don't depend on a synchronous clock. What happens if you change to a second sourced processor with slightly different timing? Keep lots of margin.

Never drive a critical signal such as clock off board without buffering. There are a very few absolutely critical signals in any system that must be noise-free. Examine your design and determine what these are, and take appropriate steps. Clock, of course, is the first that comes to mind. Another is ALE (Address Latch Enable), used on processors with a multiplexed address/data bus. A tiny bit of noise on ALE can cause your address register to latch in the middle of a data cycle, driving an incorrect address to the memories.

OK—so now your voltage levels are right. Go back to the data sheet and make sure the clock's timing is in spec.

The 8088 requires a 33% clock duty cycle. Sure, it's a little odd, but this is a fundamental rule of nature to 8088 designers. Other chips have tight duty cycle requirements as well.

Rise and fall times are just as important, though difficult to design for. Some chips have *minimum* rise/fall time requirements! It's awfully hard to predict the rise/fall time for a track routed all over the board. That's one attraction of microprocessors with a clock-out signal. Provide a decent clock-input to the chip, connect nothing to this line other than the processor, and then drive clock-out all over the board.

Motorola's 68HC16 pulls a really neat trick. You can use a 32,768-Hz standard watch crystal to clock the device. An internal PLL multiplies this to 16 MHz or whatever, and drives a clock output to feed to the rest of the board. This gets around many of the clock problems and gives a "free" accurate time-of-day clock source.

Reset

The processor's reset input is another source of trouble. Like clock, some processors have unusual input voltage requirements for reset. Be wary.

Other chips require synchronous circuits. The old Z280 had a very odd timing spec, clearly spelled out in the documentation, that everyone ignored only to find massive troubles getting the CPU to start. I think every single Z280 design in the world suffered from this particular ill at one time or another.

Sometimes slew rate is an issue. The old RC startup circuit generates a long ramp that some processors cannot tolerate. You might want to feed it into a circuit with hysteresis, like a Schmidt Trigger, to clean up the ramp.

The more complex CPUs require a long time after power-up to stabilize their internal logic. Reset cannot be unasserted until this interval goes by. Further complicating this is the ramp-up time of the system power supply, as the CPU will not start its power-up sequence until the supply is at some predefined level. The 386, for example, requires 2^{19} clock cycles if the self-test is initiated before it is ready to run.

Think about it: in a 386 system four events are happening at once. The power supply is coming up. The CPU is starting its internal power-up sequence. The clock chip is still stabilizing. The reset circuit is getting ready to unassert reset. How do you guarantee that everything happens to spec?

The solution is a long time delay on reset, using a circuit that doesn't start timing out until the power supply is stable. Motorola, Dallas, and others sell wonderful little reset devices that clamp until the supply hits 4.5 volts or so. Use these in conjunction with a long time constant so the processor, power supply, and clocks are all stable before reset is released.

When Intel released the 188XL they subtly changed the timing requirements of reset from that of the 188. Many embedded systems didn't function with this "compatible" part simply because they weren't compliant with the new chip's reset spec. The easy solution is a three-pin reset clamp.

The moral? Always read the data sheets. Don't skip over the electrical specifications with a mighty yawn. Those details make the difference between a reliable production product and a life of chasing mysterious failures.

One of my favorite bumper stickers reads "Question Authority." It's a noble sentiment in almost all phases of life . . . but not in designing embedded systems. Obey the specifications listed in the chip vendors' datasheets!

If you've read many annual reports from publicly held companies, you know that the real meat of their condition is contained in the notes. This is just as true in a chip's data sheet. It seems no one specifies sink and source current for a microprocessor's output, but the specification of the device's Vol and Voh will always reference a note that gives the test condition. This is generally a safe maximum rating.

With watchdog timers and other circuits connected to reset inputs, be wary of small timing spikes. I spent several frustrating days working with an AMD part that sometimes powered up oddly, running most instructions fine but crashing on others. The culprit was a subnanosecond spike on the reset input, one too fast to see on a 100-MHz scope.

Homemade battery-backed-up SRAM circuits often contain reset-related design flaws. The battery should take over, maintaining a small bias to the RAM's Vcc pins, when main power fails. That's not enough to avoid corrupting the memory's contents, though.

As power starts to ramp down, the processor may run crazy for a while, possibly creating errant writes that destroy vast amounts of carefully preserved data in the RAM. The solution is to clamp the chip's reset input *as soon as power falls below the part's minimum Vcc* (typically 4.75 volts on a 5-volt part).

With reset properly asserted, Vcc now at zero, and the battery providing a bit of RAM support, be sure that the chip select and write lines to the RAM are in guaranteed "idle" states. You may have to use a small pull-up resistor tied to the battery, but be wary of discharging the battery through the resistor when the system is operating normally.

And be sure you can actually pull the line up despite the fact that the driver will experience Vcc's from +5 to zero as power fails. The cleanest solution is to avoid the problem entirely by using a RAM with an active high chip select, which you clamp to zero as soon as Vcc falls out of spec.

Despite our apparent digital world, the harsh reality is that every component we use pushes electrons around. Electrical specifications are every bit as important to us as to an analog designer. This field is still electronic engineering filled with all of the tradeoffs associated with building things electronic. Ignore those who would have you believe that designing an embedded system is nothing more than slapping logic blocks together.

Small CPUs

Shhhh! Listen to the hum. That's the sound of the incessant information processing that subtly surrounds us, that keeps us warm, washes our clothes, cycles water to the lawn, and generally makes life a little more tolerable. It's so quiet and keeps such a low profile that even embedded designers forget how much our lives are dominated by data processing. Sure, we rail at the banks' mainframes for messing up a credit report while the fridge kicks into auto-defrost and the microwave spits out another meal.

The average house has some 40 to 50 microprocessors embedded in appliances. There's neither central control nor networking: each quietly goes about its business, ably taking care of just one little function. This is distributed processing at its best.

Billions and billions of 4- to 16-bit micros find their way into our lives every year, yet mostly we hear of the few tens of millions that reside on our desktops.

Now, I'd never give up that zillion-MIP little beauty I'm hunched over at the moment. We all crave more horsepower to deal with Microsoft's latest cycle-consuming application. I'm just getting tired of 32-bit hype for embedded applications. Perhaps that 747 display controller or laser printer needs the power. Surely, though, the vast majority of applications do not.

A 4-bit controller that formed the basis for a calculator started this industry, and in many ways we still use tiny processors in these minimal applications. That is as it should be: use appropriate technology for the job at hand.

Derivatives of some of the earliest embedded CPUs still dominate the market. Motorola's 6805 is a scaled up 6800 which competed with the 8080 back in the embedded Dark Ages. The 8051 and its variants are based on the almost 20-year-old 8048.

8051s, in particular, have been the glue of this industry, corresponding to the analog world's old 741 op amp or the 555 timer. You find them *everywhere*. Their price, availability, and on-board EPROM made them the natural choice for applications requiring anywhere from just a hint of computing power to fairly substantial controllers with limited user interfaces.

Now various vendors have migrated this architecture to the 16-bit world. I can't help but wonder if this makes sense, as scaling a CPU, while maintaining backward compatibility, drags lots of unpleasant baggage along. Applications written in assembly may benefit from the increased horsepower; those coded in C may find that changing processor families buys the most bang for the buck.

Microchip, Atmel, and others understand that the volume part of the embedded industry comes from tiny little CPUs scattered with reckless abandon into every corner of the world. These are cool parts! The smaller members offer a minimum amount of compute capability that is ideal for simple, cost-sensitive systems. Higher-end versions are well suited for more complicated control applications.

Designers seem to view these CPUs as something other than computers. "Oh, yeah, we tossed in a couple of PIC16s to handle the microswitches," the engineer relates, as if the part were nothing more than a PAL. This is a bit different from the bloodied, battered look you'll get from the haggard designer trying to ship a 68030-based controller. The microcontroller is easy to use simply because it is stuffed into easy applications.

L.A. Gear sells sneakers that blink an LED when you walk. A PIC16C5x powers these for months or years without any need to replace the battery. Scientists tag animals in the wild with expendable subcutaneous tracking devices powered by these parts. In Chapter 4 I mentioned the benefit of adding small CPUs just to partition the code. There are other compelling reasons as well.

A friend developing instruments based on a 32-bit CPU discovered that his PLDs don't always properly recover from brown-out conditions. He stuffed a $2 controller on the board to properly sequence the PLD's reset signals, ensuring recovery from low-voltage spikes. The part cost virtually nothing, required no more than a handful of lines of code, and occupied the board space of a small DIP. Though it may seem weird to use a full computer for this trivial function, it's cheaper than a PAL.

Not that there's anything wrong with PALs. Nothing is faster or better at dealing with complex combinatorial logic. Modern super-fast versions are cheap (we pay $12 in singles for a 7-nanosecond 22V10) and easy to use, and their reprogrammability is a great savior of designs that aren't quite right. PALs, though, are terrible at handling anything other than simple sequential logic. The limited number of registers and clocking options means you can't use them for complicated decision making. PLDs are better, but when speed is not critical a computer chip might be the simplest way to go.

As the industry matures, lots of parts we depend on become obsolete. One acquaintance found the UART his company depended on no longer available. He built a replacement in a PIC16C74, which was pin-compatible with the original UART, saving the company expensive redesigns.

In the good old days of microcomputing, hardware engineers also wrote and debugged all of the system's code. Most systems were small enough that a single, knowledgeable designer could take the project from conception to final product. In the realm of small, tractable problems like those just described, this is still the case. Nothing measures up to the pride of being solely responsible for a successful product; I can imagine how the designer's eyes must light up when he sees legions of kids skipping down the sidewalk flashing their L.A. Gears at the crowds.

Part of the recent success of these parts comes from the aggressive use of Flash and One-Time Programmable (OTP) program memory. OTP memory is simply good old-fashioned EPROM, though the parts come without an erasure window. That small quartz opening typical of EPROMs and many PLDs is very expensive to manufacture. You can program the memory on any conventional device programmer, but, since there's no window, you can never erase it. When it's time to change the code, you'll toss the part out.

Intel sold OTP versions of their EPROMs many years ago, but they never caught on. A system that uses discrete memory devices—RAM, ROM, and the like—has intrinsically higher costs than one based on a microcontroller. In a system with $100 of parts, the extra dollar or two needed to use erasable EPROMs (which are very forgiving of mistakes) is small.

The dynamics are a bit different with a minimal system. If the entire computer is contained in a $2 part, adding a buck for a window is a huge cost hit. OTP starts to make quite a bit of sense, assuming your code will be stable.

This is not to diminish Flash memory, which has all of the benefits of OTP, though sometimes with a bit more cost.

Using either technology, the code *can* be cast in concrete in small applications, since the entire program might require only tens to hundreds of statements. Though I have to plead guilty to one or two disasters where it seemed there were more bugs than lines of code, a program this small, once debugged and thoroughly tested, holds little chance of an obscure bug. The risk of going with OTP is pretty small.

You can't pick up a magazine without reading about "time to market." Managers want to shrink development times to zero. One obvious solution is to replace masked ROMs with their OTP equivalents, as producing a processor with the code permanently engraved in a metalization layer takes months . . . and suffers from the same risk factors as does OTP. The masked part might be a bit cheaper in high volumes, but this price advantage doesn't help much if you can't ship while waiting for parts to come in.

Part of the art of managing a business is to preserve your options as long as possible. Stuff happens. You can't predict everything. Given options, even at the last minute, you have the flexibility to adapt to problems and changing markets. For example, some companies ship multiple versions of a product, differing only in the code. A Flash or OTP part lets them make a last-minute decision, on the production floor, about how many of a particular widget to build. If you have a half million dollars tied up in inventory of masked parts, your options are awfully limited.

Part of the 8051's success came from the wide variety of parts available. You could get EPROM or masked versions of the same part. Low-volume applications always took advantage of the EPROM version. OTP reduces the costs of the parts significantly, even when you're only building a handful.

Microcontrollers do pose special challenges for designers. Since a typical part is bounded by nothing more than I/O pins, it's hard to see what's going on inside. Nohau, Metalink, and others have made a great living producing tools designed specifically to peer inside of these devices, giving the user a sort of window into his usually closed system.

Now, though, as the price of controllers slides toward zero and the devices are hence used in truly minimal applications, I hear more and more from people who get by without tools of any sort. While it's hard to condone shortchanging your efficiency to save a few dollars, it's equally hard to argue that a 50-line program needs much help. You can probably eyeball it to perfection on the first or second iteration. Again, appropriate technology is the watchword; 5000 lines of assembly language on a 6805 will force you to buy decent debuggers . . . and, I'd hope, a C compiler.

You can often bring up a microcontroller-based design without a logic analyzer, since there's no bus to watch. Some people even replace the scope with nothing more than a logic probe.

An army of tool vendors supply very low-cost solutions to deal with the particular problems posed by microcontrollers. You have options—lots of them—when using any reasonable controller—far more than if you decide to embed a SPARC into your system.

Some companies cater especially to the low end. Most do a great job, despite the low cost. I recently looked at Byte Craft's array of compilers for microcontrollers from Microchip, Motorola, and National. Despite the limited address spaces of some of these parts, it's clear a decent C compiler can produce very efficient code.

One friend cross-develops his microcontroller code on a PC. Using C frees him from most processor dependencies; compile-time switches select between the PC's timer/UART, etc., and that contained in the controller. He manages to debug more than 80% of the code with no target hardware.

Working in a shop using mostly midrange processors, I'm amazed at the amount of fancy equipment we rely on, and am sometimes a bit wistful for those days of operating out of a garage with not much more than a soldering iron, a logic probe, and a thinking cap. Clearly, the vibrant action in the controller market means that even small, under- or uncapitalized businesses still can come out with competitive products.

Watchdog Timers

I'm constantly astonished by the utter reliability of computers. While people complain and fume about various PC crashes and other frustrations, we forget that the machine executes millions of instructions per second, even when sitting in an idle loop. Smaller device geometries mean that sometimes only a handful of electrons represent a one or zero. A single-bit failure, for a fleetingly transient bit of time, is disaster.

Yet these failures and glitches are exceedingly rare. Our embedded systems, and even our desktop computers, switch trillions of bits without the slightest problem.

Problems can and do occur, though, due more often to hardware or software design flaws than to glitches. A watchdog timer (WDT) is a good defense for all but the smallest of embedded systems. It's a mechanism that restarts the program if the software runs amok.

The WDT usually resets the processor once every few hundred milliseconds unless reset. It's up to the firmware to reinitialize the watchdog timer, restarting the timing interval. The code tickles the timer frequently, restarting the countdown interval. A code crash means the timer counts down without interruption; at time-out, hardware resets the CPU, ideally bringing the system back on-line.

The first rule of watchdog design is to drive the CPU's reset input, not an interrupt (such as NMI). A WDT time-out means that something awful happened, something that may have left the CPU in an unpredictable scrambled state. Only RESET is guaranteed to bring the part back on-line.

The non-maskable interrupt is seductive to some designers, especially when the pin is unused and there's a chance to save a few gates. For better or worse, NMI—and all other interrupt inputs—is not fail-safe. Confused internal logic will shut down NMI response on some CPUs.

On other chips a simple software problem can render the non-maskable interrupt unusable. The 68K, for example, will crash if the stack pointer assumes an odd value. If you rely on the WDT to save the day, driving an interrupt while SP is odd results in a double bus fault, which puts the CPU in a dead state until it's reset.

Next, think through the litigation potential of your system. Life-threatening failure modes mean you've got to beware of simple watchdog timers! If a single I/O instruction successfully keeps the WDT alive, then there's a real chance that the code might crash but continue to tickle the timer. Some companies (Toshiba, for example) require a more complex sequence of commands to the timer; it's equally easy to create a PLD yourself that requires a fiendishly complex WDT sequence.

It's also a very bad idea to put the WDT reset code inside of an interrupt service routine. It's always intriguing, while debugging, to find your code crashed but one or more ISRs still functioning. Perhaps the serial receive routine still accepts characters and echoes them to the sender. After all, the ISR by definition runs independently of the rest of the code, so will often continue to function when other routines die. If your WDT tickler stays alive as the world collapses around the rest of the code, then the watchdog serves no useful purpose.

This problem multiplies in a system with an RTOS, as a reliable watchdog monitors *all* of the tasks. If some of the tasks die but others stay alive—perhaps tickling the WDT—then the system's operation is at best degraded.

In this case write the WDT code as its own task, driven by a timer. All other tasks send messages to the watchdog process, indicating "I'm alive." Only when the WDT activity sees that all tasks that should have checked in are indeed operating does it service the watchdog. If you use RTOS-supplied messaging to communicate the tasks' health—rather than dreaded though easy global variables—there's little chance that errant code overwriting RAM can create a false indication that all's OK.

Suppose the WDT does indeed find a fault and resets the CPU. Then what? A simple reset and restart may not be safe or wise.

One system uses very high-energy gamma rays to measure the thickness of steel. A hardware problem led to a series of watchdog time-outs. I watched, aghast, as this system cycled through WDT resets about once a second, *each time opening the safety shield around the gamma ray source!* The technicians were understandably afraid to approach close enough to yank the power cord.

If you cannot guarantee that the system will be safe after the watchdog fires, then you simply must add hardware to put it in a reasonable, non-dangerous, mode.

Even units that have no safety issues suffer from poorly thought-out WDT designs. A sensor company complained that their products were getting slower. Over time, and with several thousand units in the field, response time to user inputs degraded noticeably. A bit of research showed that their system's watchdog properly drove the CPU's reset signal, and the code then recognized a warm boot, going directly to the application with no indication to the users that the time-out had occurred. We tracked the problem down to a floating input on the CPU that caused the software to crash—up to several thousand times per second. The processor was spending most of its time resetting, leading to apparently slow user response.

If your system recovers automatically from a WDT time-out, add an LED or status display so users—or at least the programmers!—know that the system had an unexpected reset. Don't use a bit of clever watchdog code to compensate for software or hardware glitches.

> Should embedded systems have a reset switch?
>
> It seems almost traditional to put a reset switch on the back panel of an embedded system. When something horrible happens, hit the reset and retry! Doesn't this make the customer feel that we don't trust our own products? Electronic systems never had reset switches until the introduction of the microprocessor. Why add them now?
>
> A reset switch is no substitute for flaky hardware. It's pretty easy (or, at least possible) to design robust, reliable microprocessor circuits. Any failure is most likely to be a hard fault that a simple reset will not cure.
>
> This argument implies that a reset switch is mostly useful to cure software bugs. We have a choice of writing 100% reliable code or adding some sort of an escape hatch for the user. I hereby proclaim, "We shall all now write correct code."
>
> The problem is now cured.
>
> OK, so perhaps a bug just might creep in once in a while. My feeling is that a reset switch is still a mistake. It conveys the message that no one really trusts the product. It's much better to include a very robust watchdog timer that asserts a good, hard reset when things fall apart. The code might still be unreliable, but at least we're not announcing to the world that bugs are perhaps rampant. Remember when Microsoft eliminated the Unexpected Application Error message from Windows 3.1 . . . by renaming it?
>
> No watchdog is perfect, but even a simple one will catch 99% of all possible code crashes. Combine this percentage with the (ideally) low probability of a software crash, and the watchdog failure rate falls to essentially zero.

Making PCBs

In the bad old days we created wire-wrapped prototypes because they were faster to make than a PCB, and a lot cheaper. This is no longer the case. Except for the very smallest boards, the cost of labor is so high that it's hard to get a wire-wrapped prototype made for less than $500 to several thousand dollars. Turnaround time is easily a week.

Cheap autorouting software means any engineer can design a PCB in a matter of a couple of days—and you'll have to do this eventually anyway, so it's not wasted time. Dozens of outfits will convert your design to a couple of PCBs in under a week for a very reasonable price. How much? Figure $1000–1500 for a 50-square-inch 4- to 6-layer board, with one-week turnaround.

It's magic. Modem your board design to the vendor, and days later FedEx delivers your custom design, ready for assembly and test.

PCBs are much quieter, electrically, than their wire-wrapped brethren. With fast rise times and high clock rates, noise is a significant problem even in small embedded designs. I've seen far too many cases of "Well, it doesn't work reliably, but that's probably due to the wire wrap. It'll probably get better when we go to PC." These are clearly cases where the prototype does not accomplish its prime objective: identify and fix all risk factors.

Always build your prototype on a PCB, never on wirewrap or other impedance-challenged technologies. And figure on using a multilayer design, with unadulterated power and ground planes. Modern logic is just too fast, too noisy, and too intolerant of ground bounce and other impedance issues to try and mix power and signals on any PCB layer.

The best source for information about speed and noise issues on PC boards is *High Speed Digital Design—A Handbook of Black Magic*, by Howard Johnson and Martin Graham (1993, PTR Prentice Hall, NJ). This is a must-read for all digital engineers. If you felt that your college electromagnetics was a flunk-out course, one you squeaked through, fear not. The authors do use plenty of math, but their prose descriptions are so lucid you'll gain a lot of insight by just reading the words and skipping over the equations.

Design your prototype PCB with room for mistakes. Designing a pure surface-mount board? These usually use tiny vias (the holes between layers) to increase the density. *Think* about what happens during the prototyping phase: you'll make design changes, inevitably implemented by a maze of wires. It's impossible to run insulated wire through the tiny holes! Be sure to position a number of unusually large vias (say, 0.031") around the board that can act as wiring channels between the component and circuit sides of the board.

Add pads for extra chips; there's a good chance you'll have to squeeze another PAL in somewhere. My latest design was so bad I had to glue on five extra chips. Guess who felt like an idiot for a few days. . . .

Always build at least two copies of each prototype PCB. One may lag the other in engineering modifications, but you'll have options if (when) the first board smokes. Anyone who has been at this for a while has blown up a board or two.

I generally buy three blank prototype PCBs, assemble two, and use the third to see where tracks run. Though sometimes you'll have to go back to the artwork to find inner tracks, it sure is handy to have the spare blank board on the bench during debug.

> It's scary how often the firmware group receives a piece of "functional" prototype hardware from the designers accompanied by nothing more than the schematics—schematics that are usually incomprehensible to the software folks, made even more abstruse by massive use of PLDs and similar functional blocks plopped down on the page, with perhaps hundreds of connections. They are documentation black holes—every signal goes in, and presumably something comes out, but without the designer's suite of design tools even the brightest firmware person will never make sense of the design.
>
> Where does one draw the line between the responsibilities of the hardware designers and those of the firmware folks? Should the designers include device drivers? Seems reasonable to me, since surely they did indeed at least hack together a bit of code to test each device. Why not structure the development plan to make this test code part of the framework of the final software? The hardware tends to be so complex now that it's unfair to give "naked iron" to the software people. At the very least, deliver low-level drivers with well-defined interfaces.
>
> If you live and breathe hardware only, do talk to your software counterparts. You may be surprised to learn that all too often your cool new product makes debugging the code practically impossible. Poor design decisions might seriously affect the firmware schedule. All embedded people must understand that their creation does not exist in isolation; the code and the chips all function together, to form the seamless gestalt that (you hope) delights the user.

Changing PCBs

After spending a couple of months writing code, it's a bit of a shock to come back to the hardware world. Fixing bugs is a real pain! Instead of a quick edit/compile, you've got to break out a soldering iron, wire, parts, and then manipulate a pin that might be barely visible.

PALs, FPGAs, and PLDs all ease this process to some extent. Many changes are not much more difficult than editing and recompiling a file. It is important to have the right tools available: your frustration level will skyrocket if the PAL burner is not right at the bench.

FPGAs that are programmed at boot time via a ROM download usually have a debugging mechanism—a serial connection from the device to your PC, so you can develop the logic in a manner analogous to using a ROM emulator. Be sure to put the special connector on your design, and buy the little adapter and cable. Burning ROMs on each iteration is a terrible waste of time.

PLDs often come like EPROMs, in ceramic packages with quartz erasure windows. These are great . . . if you were clever enough either to socket the parts, or to have left room around the part for a socket.

On through-hole designs I generally have the technicians load sockets for every part on the prototype. I want to replace suspected failed devices quickly, without spending a lot of time agonizing over "Is it really dead?"

Sockets also greatly ease making circuit modification. With an 8-layer board it's awfully hard to know where to cut a track that snakes between layers and under components. Instead, remove the pin from the socket and wire directly to it.

You can't lift pins on programmable parts, as the device programmer needs all of them inserted when reburning the equations. Instead, stack sockets. Insert a spare socket between the part and the socket soldered on the board. Bend the pins up on this one. All too often the metal on the upper socket will, despite the bent-out pin, still short to the socket on the bottom. Squish the metal in the bottom socket down into the plastic to eliminate this hard-to-find problem.

Surface-mount parts are much more problematic. Get a good set of dental tools and a very fine soldering iron, so you can pry up pins as needed. You'll need a bright light with magnifier, a steady hand, and abstinence from coffee. A decent surface-mount rework machine (such as from Pace Electronics) is essential; get one that vectors hot air around the IC's pins. Don't even try to use conventional solder on fine-pitch parts; use solder paste instead, and keep it fresh (usually it's best stored in a fridge).

Since SMT is so tough, I always make prototype boards with tracks on the outer layers. Sure, the final version might reverse this (power and ground outside to reduce emissions), but reverse the layering during debug. It's easy to cut tracks with an X-Acto knife.

Every engineer needs at least two X-Acto knives. One is for fingernail cleaning, cutting open envelopes, and tossing at the dartboard. The other is only for PCB work and always has a new, sharp blade. Keep 50 or 100 spare blades in your drawer, since PCB work invariably breaks the very sharp and very essential pointy end off in no time.

Planning

Engineers have managers, who "run" projects, ensuring that resources are available when needed, negotiate deadlines and priorities with higher-ups, and guide/mentor the developers toward producing a decent product on time. Planning is one of any manager's main goals. Too often, though, managers do planning that more properly belongs to the engineers. You know more about what your project needs than your boss ever will; it's silly, and unfair, to expect him to deal with all of the details.

There are many great justifications for a project running late. In engineering it's usually impossible to predict all of the technical problems you'll encounter! However, lousy planning is simply an unacceptable, though all too common, reason.

I think engineers spend too much time *doing*, and not enough time *thinking about doing*. Try spending two hours every Monday morning planning the next week and the next month. What projects will you be working on? What's their status? *What is the most important thing you need to do to get the projects done?* Focus on the desired goal, and figure out what you need to do to get there. Do you need to order parts? Tools? Does some of your test equipment need repair or calibration?

Find the critical paths and do what's required to clear the road ahead. Few engineers do this effectively; learn how, and you'll be in much higher demand.

When you're developing a rush project (all projects are rush projects . . .), the first design step is a block diagram of the each board. From this you'll create the schematic, then do a PCB layout, create a bill of materials, and finally, order parts for the prototype.

Not. The worst thing you can do is have a very expensive quick-turn PCB arrive, with all of the components still on back order. The technicians will snicker about your "hurry up and wait" approach, and management will be less than thrilled to spend heavily for fast-turn boards that idle away the weeks on a shelf.

Buy the parts first, before your design is complete. Surely you'll know what all of the esoteric parts are—the CPU, odd analog components, sensors, and the like. These are likely to be the hardest and slowest to get, so put them on order immediately.

The nickel and dime components, such as gates and PALs, resistors and capacitors, are hard to pin down until the schematic is complete. These should mostly be in your engineering spares closet. Again, part of planning is making sure your lab has the basic stuff needed for doing the job, from soldering irons to engineering spares. Make sure you have a good selection of the sort of components your company regularly uses, and avoid the temptation to use new parts unless there's a good reason.

CHAPTER **7**

Troubleshooting Tools

Developers expect long, painful debugging sessions. We plunge into system debug without thinking through the benefits and perils of this step, and as a result generally wind up in a nightmare of bugs and schedule panics.

As discussed in Chapter 2, a careful program of Code Inspections will eliminate 70 to 80% of the bugs in a system before the first bit of testing commences. The same chapter also shows how a careful developer can count and manage bugs to identify bad code and take appropriate action early.

An HP study concluded that the debugging process itself is flawed, as it generally exercises only half of the code. That is, no one is smart enough to construct a test that checks every possible IF-THEN condition, each CASE in a SWITCH statement. This surely reinforces the need for Code Inspections, but clearly even Inspections combined with test will result in substantial chunks of untested—and thus buggy—code.

> The math is simple. Most code runs around a 5% bug rate after compiler-found syntax errors are corrected. A little 10,000-line program will typically have about 500 bugs before inspection and test. Code Inspections will identify about 70 to 80% of these, leaving some 100 still latent. Test, then, is our last defense against shipping a bug-ridden product . . . but test only exercises half the code, leaving 50 bugs still in the finished unit!

This is clearly unacceptable. There are a few solutions:

1. Single-step though all of the code. Keep a listing handy, on paper, and check off each branch and decision node as you step through it, running tests until every bit of code has been executed. The downside of this, of course, is that single-stepping destroys the real-time nature of most embedded systems.
2. Construct tests guaranteed to run through every decision node. This means modifying the test procedure after you've written the firmware to ensure that the tests are robust enough to run through every node.
3. Buy a fancy tool. Applied Microsystems and HP both make code coverage tools that identify unexecuted lines of code, watching system operation in real time. These tools serve as a complement to option 2, as you'll still have to construct appropriate tests. Still, if bugs are unacceptable, then the fancy tools are probably necessary to ensure quality.

No management techniques or methodologies will ever eliminate the need for test and debug. The late, great Deming taught the world that it's impossible to test quality into a system; quality is a characteristic of the design, not of our ability to find and fix bugs. Yet no matter how elegant the design, test is always important, always a crucial validation of the code.

Tools

Your lovingly crafted, finely tuned masterpiece of engineering will not work. Period. Sometimes it's a little frightening when we discover the real scope of our errors in a design. How often have you thought, in a bleak moment of despair, "I'll *never* make this stupid thing work!"

But that's why we build prototypes. Prototypes are not expected to work at first. Electronics engineering is perhaps one of the last great areas where we can and should build test systems that are meant to be thrown away once their contribution to the design process is done.

Although this is no excuse for doing a sloppy job of design, *expect* problems. Develop an engineering strategy that expects problems as part of the design process, rather as a reaction to (surprise!) a mistake. Set up a system where you extract every bit of meaning from problems and their eventual solutions. Don't be like the engineer who finds a mistake, cuts and pastes a repair . . . and then forgets to document it, dooming himself or some other poor soul to troubleshooting the same symptom all over again.

Above all, don't plunge into the troubleshooting madness too quickly. Debugging some embedded projects can take months. Invest time up front to organize your workbench, acquire the tools, and learn to use them effectively.

Who built the first lathe? The first oscilloscope? It's hard to conceive how these pioneers bootstrapped their efforts, somehow breaking the cycle of needing equipment X to produce equipment X. Though this surely proves that modern tools are dispensable, only a fool would wish to repeat the designers' Herculean efforts.

Select and buy a tool for one reason only: to save time! Since this is a rapidly evolving field, expect to continuously invest in new equipment that keeps you maximally productive. Surely no one would advocate using 286 computers in a Pentium world, yet far too many companies sentence their engineers to hard labor by refusing to upgrade scopes, compilers, and emulators when advancing technology obsoletes the old.

Every bookstore is crammed with volumes of sage advice for getting more from each hour. Never forget that the fundamental rule of time management is to work smart; in the computer business, delegate as much as possible to your electronic servants that cost so little compared to an engineer's salary.

Debuggers—of every ilk—do one fundamental thing: provide visibility into your system. Features vary, but all we ask of a debugger is, "Tell me what is going on." Sometimes we're interested in procedural flow (single-stepping, breakpointing); other times it's function timing or dependencies or memory allocation. Regardless, we simply expect our tools to reveal hidden system behavior. Only after we *see* what's going on can we use our brains to understand "why that happened," and then apply a fix.

Before talking about specific tools, let's look at the features we'd like to see in any sort of debugger (see Figure 7-1), and only then see how the tools match feature requirements.

Source-level debugging—If you write in C, debug in C. There is no more important feature than an environment that lets you debug in the same context in which you originally wrote the code. If the debugging tools won't automatically call up the appropriate source files showing where the current program counter lies, then count on long, painful days of despair trying to make things work.

Tools, after all, are the intelligent assistants that provide us a level of abstraction between the awful bits and bytes the computer uses and our code. The source-level debugger is the critical ingredient that connects us to the

Feature	Emulator	BDM	ROM monitor	Logic analyzer	ROM emulator
Source debugging	Yes	Yes	Yes	Some	Yes
Download code	Yes	Yes	Yes	No	Yes
Single-step	Yes	Yes	Yes	No	Yes
Basic breakpoints	Yes	Yes	Yes	No	Yes
Display/alter registers et al.	Yes	Yes	Yes	Yes	Yes
Watch variables	Yes	Yes	Yes	Yes	Yes
Real-time trace	Yes	No	No	Yes	No
Event triggers	Yes	No	No	Yes	No
Overlay RAM	Yes	No	No	No	Yes
Shadow RAM	Some	No	No	No	No
Hardware breakpoints	Yes	Some	No	No	Some
Complex breakpoints	Yes	No	No	Yes	No
Time stamps	Yes	No	No	Yes	No
Execution timers	Yes	No	No	Yes	No
Nonintrusive access	Yes	Yes	No	Yes	No
Cost	Very high	Cheap	Cheap	High	Cheap

FIGURE 7-1 Typical features of debugging tools.

tool itself (emulator, ROM monitor, etc.) and our original source code. Hit a breakpoint, and the debugger will highlight the current address in the current source file. You view your original source code with comments. The debugger shows data items in their native type (ints as decimal integers, floats as floating-point numbers, strings as ASCII text), not as raw, impossible-to-decipher hex codes.

The source-level debugger is a program that runs on the PC and that communicates with the emulator or whatever. It's an essential part of a professional debug environment.

If your toolchain won't include a decent source debugger, triple your debugging time, since most of your effort will be spent in the unrewarding (and, frankly, stupid) task of correlating bits and bytes to source code.

Nonintrusive access—Nonintrusive access means the tool "gets inside the head" of your target system without consuming the target's memory, peripherals, or any other resources.

As CPUs get more complex, though, all tools have more restrictions that you, the user, must understand. If the part has cache, will the tool work with cache enabled? A more insidious—and common—problem stems from pins shared between several functions. If address line 18, for example, can be changed to a timer output under program control, will the emulator gork? Call the vendor and ask for the "restriction list" before buying any debugging tool.

Real-time trace—Trace captures the execution stream of your code in real time, displaying it in the original C or C++ source. Trace depths are measured in frames, where one frame is one memory or I/O transaction—thus, a single instruction may eat up several frames of storage.

Trace width is given in bits, and generally includes the address, data, and some of the control busses, perhaps also with external inputs (to show how the code and hardware synchronize), and timing information. Widths vary from 32 bits to more than 100.

Trace is most useful for capturing real-time code—such as the execution of an ISR—without slowing the system at all. It's generally nonintrusive.

Trace is mostly associated with logic analyzers and emulators. Be aware that as CPUs get more complex, many emulators capture only the address bus in the trace buffer . . . which means you'll have no view of the data transactions associated with the code.

Event triggers and filters—Event triggers start and stop trace acquisition. You define a condition (say, "when foobar = 23"); in real time the tool detects that condition and starts/stops the trace collection. Filters include or exclude cycles from the trace buffer (it makes little sense, for example, to acquire the execution of a delay routine).

Even with the hundreds of thousands of trace frames offered by some devices, there's never enough depth to collect more than a tiny bit of the code's operation. Triggers and filters let you specify exactly what gets captured. The skillful use of triggers and filters reduces your need for deep trace and greatly reduces the amount of acquired data you'll have to sift through.

Overlay RAM—also known as emulation RAM—though physically inside of an emulator, is mapped into the target processor's address space. Overlay RAM replaces the ROM or Flash on your system so you can quickly download updated code as bugs are discovered and repaired. ICEs provide great latitude in mapping this RAM, so you can change between the emulator's memory and target memory with fine granularity. A singular benefit of overlay is that you can often start testing your code before the target hardware is available.

Today's Flash-based systems might seem to eliminate the need for overlay, but in fact Flash programs more slowly than RAM, leading to longer download times.

Shadow RAM—When the emulator updates the source debugger's windows, it interrupts the execution of your code to extract data from registers, I/O, and memory—an interruption that can take from microseconds to milliseconds. Shadow RAM is a duplicate address space that contains a current image of your data that the tool can access without interrupting target operation.

Hardware breakpoints—Breakpoints stop program execution at a defined address, without corrupting the CPU's context. A software breakpoint replaces the instruction at the breakpoint address with a one byte/word "call." There's no hardware cost, so most debuggers implement hundreds or thousands. Hardware breakpoints are those implemented in the tool's logic, often with a big RAM array that mirrors the target processor's address space. Hardware breakpoints don't change the target code; thus, they work even when you're debugging firmware burned in ROM.

Some pathological algorithms defy debugging with software breakpoints. A ROM test routine, for example, might CRC the code itself; if the debugger changes the code for the sake of the breakpoint, the CRC will fail. There's no such restriction with a hardware breakpoint.

Hardware breakpoints do come at a cost, though, so some tools offer lots of breakpoints, with a few implemented in hardware and the bulk in software.

Complex breakpoints—Simple BPs stop the program only on an instruction fetch ("stop when line 124 is fetched"). Their *complex* cousins, though, halt execution on data accesses ("stop when 1234 is written to foobar"). They'll also allow some number of nested levels ("stop when routine activate_led occurs after led_off called"). Though some tools offer quite a diverse mix of nesting levels, few developers ever use more than two.

Desktop debuggers such as that supplied with Microsoft's VC++ usually offer complex breakpoints—but they do not run in real time, and they impose significant performance penalties. Part of the cost of an ICE is in the hardware required to do breakpoints in real time.

It's important to understand that a simple hardware or software breakpoint stops your code *before* the instruction is executed. Complex BPs, especially when set on data accesses, stop execution *after* the instruction completes. On processors with prefetchers it's not unusual for the complex breakpoint to skid a bit, stopping execution several instructions later.

Time stamping—Emulators and logic analyzers often include time information in the trace buffer. Time stamps usually eat up about 32 bits of trace width. Combined with the trace system's triggers, it's easy to perform quite involved timing measurements.

Emulators

In-Circuit Emulators (ICEs) have always been the choice weapons in the war on bugs. Yet, for as long as I can remember pundits have been predicting their death. Though it seems as quaint as IBM's 1950s prediction that the worldwide market for computers was merely a couple of dozen, in fact 20 years ago many people believed that the 4-MHz Z80 would spell doom for ICEs. "4 MHz is just too fast," they proclaimed. "No one can run those speedy signals down a cable."

Time proved them wrong, of course. Today's units run at 60+ MHz on processors with single-clock memory cycles, an astonishing achievement.

Is an end yet in sight? I believe so, though the limiting frequency is a bit hazy. Today's approach of putting all or much of the ICE's electronics on the pod removes the cabling and bus driver problems, but electrons do move at a finite speed and even the fastest of circuits have nonzero propagation delays.

CPU vendors squeeze the last bit of clock rates from their creations partly by tuning their chips ever more exquisitely to the rest of the system's memory and I/O. Clearly, an intrusion by any sort of development tool will at best be problematic. Yes, today's Pentium emulators do work. Will tomorrow's units be able to handle the continued push into stratospheric clock rates? I have doubts.

Packages are creating another sort of problem. Heat, speed, and size constraints have yielded a proliferation of packaging styles that challenge any sort of probing for debugging. If you've ever tried to use a scope on a 208-pin PQFP device or, worse, a 100-pin TQFP, you know what I mean. Yes, some tremendously innovative probing systems exist—notably those from Emulation Technology and HP. Despite these, it's still difficult at best to establish a reliable connection between a target CPU and any sort of hardware debugger, from a voltmeter to an ICE.

Surface-mount devices have exposed pins that you at least have a prayer of getting to. Newer devices don't. The BGA (Ball Grid Array) package, which is suddenly gaining favor, connects to a PC board via hundreds of little bumps on the underside of the package—where they are completely inaccessible. Other technologies bond the silicon itself under a dab of epoxy directly to the board. All of these trends offer various system benefits; all make it difficult or impossible to troubleshoot software and hardware.

OK, you smirk, these issues only apply to the high end of the embedded market, where clock rates—and production costs—soar with the eagles. Other, subtle influences, though, are wreaking havoc on the low end.

Take microcontrollers, for example. These CPUs have ROM and RAM on-board, giving a very simple, very inexpensive one-chip solution for simple 8- and 16-bit applications. The 8051 is the classic example of this, and indeed has been an amazing success that has survived 20 years of assault by other, perhaps more capable, processors.

Single-chip solutions are tough to debug, though, since the on-board memory means there's generally no address/data bus coming to the outside world. An extreme example is Microchip's 8-pin PIC part. Eight pins!

Various debugging solutions exist, but the traditional solution is the bond-out chip, a special version of the processor, with extra pins that bring all important signals to the outside world, especially those oh-so-critical address and data lines needed to track program execution. With a proper bond-out-based ICE you can track everything the code does, in real time, with no compromises. Perfect, no?

Well, a few wrinkles are starting to surface. For one, the chip vendors *hate* making bond-outs. The market is essentially zero, yet every time the processor's mask gets revised a new bond-out is needed. In the old days chip vendors swallowed hard, but did make them reasonably available.

Now this is less common. With the 386EX (which is not a microcontroller, but which benefits from a bond-out) Intel announced that only a handful of vendors would get access to the special version of the part, probably to some extent increasing the cost of tools. Is this an indication of the beginning of the end of generally available bond-out parts?

Sometimes the bond-out is not kept to current mask revisions. I know of at least one case where a vendor provides bond-outs that will not run at full speed, essentially removing the critical visibility of real-time execution from developers. This situation puts you in the awful conundrum of deciding, "Should I buy an expensive tool . . . that forces me to run at half speed, no doubt destroying all timing relationships?"

Sometimes—often—the bond-outs will not run at reduced voltages. Your 3-volt system might require a pod that is a convoluted mix of 3- and 5-volt technologies, creating additional propagation delays as voltages get translated. In effect, a nonintrusive tool becomes subtly more intrusive, in ways that are hard to predict. Voltages are declining fast—some CPUs now run at sub-1-volt levels—so the problem can only get worse.

A very scary development is the incredible proliferation of CPUs. Vendors are proud of their ability to crank out a new chip by pressing a few buttons on a CAD system, changing the mix of peripherals and memory, producing variant number 214 in a particular processor family. Variants are a sign of a good, healthy line of parts (look at that mind-boggling array of 8051 parts), but are a nightmare for tool vendors. Each requires new hardware, software, support, evaluation boards, and the like. In the "good old days," when we saw only a few new parts per year per family, support was easy to find. Now my friends who make microcontroller tools complain of the frantic pace needed to support even a subset of the parts.

As a tool consumer you probably don't care about the woes of the vendors. But part proliferation creates a problem that hits a bit closer to home: for any specific variant there may only be a handful of customers. Tool support may never exist for that part if vendors feel there's not a big enough market. An odd fact of the tool market (from compilers to ICEs) is that the health of the market is a function of the number of customers using a chip, not the number of chips used. CPU vendors are happy to get one or two huge design wins, say an automotive company that sucks up millions of parts per year. Tool folks might only sell a couple of units to such a customer, far too few to pay their huge development costs.

Yet, despite the problems inherent with any tool so closely coupled to the CPU, the ICE is without a doubt the most powerful and most useful tool we have for debugging an embedded system. Only an ICE gives a nonintrusive real-time view of the firmware's operation.

Why use an ICE?

- If your target hardware is not perfect, most other tools will not function well. An ICE is probably the most useful tool around for finding and troubleshooting hardware as well as software problems.
- The ICE uses no target resources. In general, all ROM, RAM, and interrupts will be untouched.
- There is no better way to debug real-time code than using trace coupled with extensive triggering capabilities. The emulator captures the busses, and, in conjunction with the source-level debugger, correlates raw bus activity to your C source files.

Emulator downsides include:

- No tool is more expensive than an emulator.
- As discussed earlier, speed and mechanical issues mean that some systems will just not be candidates for emulator-based debugging.

142 THE ART OF DESIGNING EMBEDDED SYSTEMS

- ICEs can be finicky beasts to tame. With a hundred or more connections to your target hardware, the smallest bit of dirt, vibration, or bad luck can cause erratic operation that will drive your developers out of their minds. For this reason I always recommend soldering the emulator to an SMT part, rather than using a clip-on connection. Find a reliable hook-up scheme early, to avoid infinite frustration later.

BDMs

CPU cores hidden away inside ASICs give fabulously small systems, yet that buried processor is all but impossible to probe. Couple bus cycles within fractions of a nanosecond to a peripheral and you leave no margin for your tools. One-off CPUs, whether from burying a VHDL virtual processor inside a high-integration part, or from the huge explosion of derivatives of popular parts, are often tool orphans. Tool vendors, after all, won't invest huge sums in developing products for a particular CPU unless they see a large, healthy market for their offerings.

Even seemingly boring issues such as device packaging further isolate us from the processor. If we can't probe it, we can't see what's going on. We lose the visibility needed to find bugs.

The trend is to separate run control from real-time trace. "Run control" means those simple debugging features that we'd expect even in nonembedded work: simple breakpoints, single-stepping, and access to processor resources, memory, and peripherals. Probably 95% of all debugging uses nothing more than these relatively simple features. Trace, though, demands real-time access to the entire data, address, and control busses, and so is generally a rather thorny and expensive part of any emulator.

But the promise of a serial debugger remains seductive, given that just a few wires replace the hundreds of connections used by an emulator or logic analyzer. Motorola recognized this early on and created the Background Debug Mode (BDM), a feature first found on the 683xx and 68HC16 processors, since extended and incorporated on many other chips.

BDM is a bit of specialized debugging hardware built right into the chip (Figure 7-2). Transistors are so cheap it makes sense to build a debug interface into even production chips. Clearly this overcomes one major objection of bond-outs: the "stepping level" of the production IC is always identical to the debug part . . . because they are one and the same.

BDMs eliminate all speed and packaging issues. As part of the silicon, the debugger runs as fast as the chip; the interface to the outside world

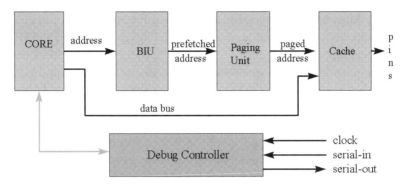

FIGURE 7-2 A BDM/JTAG debugger adds logic on the CPU itself.

is inherently not coupled to raw processor speed. Connection problems go away, since you just run a few CPU pins to a special debug connector.

Implementations vary, but a processor with BDM dedicates a few pins to a serial debugging channel (though sometimes other functions might be multiplexed onto them). Customers demand high-speed screen updates, so this is a synchronous communications scheme that includes a clock pin, supporting serial speeds beyond 1 Mbps.

Development tool vendors sell you a connection to this channel, ranging from a high-end very fast link to something no more complicated than a two-IC interface to a PC's comm port . . . and, of course, a source-level debugger. The software interfaces to your code and formats your requests to single-step or display data to meta-commands transmitted to the CPU chip (on the BDM link).

The original BDM implementation shared microcode with the processor's main execution stream. Commands processed by the debug link thus stopped normal program execution. Although this was tolerable for simple applications, users of real-time operating systems, in particular, wished to examine and alter system state without bringing the entire program to its knees. BDM+, on the ColdFire CPUs, uses a totally independent set of hardware to allow concurrent program execution and debugging.

MIPs, Intel, TI, and others provide serial debugging via various extensions of the JTAG (Joint Test Access Group) standard (IEEE 1149.1). JTAG, too, is a synchronous serial interface, one originally defined to promote testability of complex boards. Though the implementation details differ from those for BDM, in all significant user respects it offers the same sort of functionality and level of complexity.

BDM and JTAG hardware on board the processor can't waste transistors, as ultimately increasing the chip's complexity drives the cost of the part up. Most implementations, therefore, rely on software rather than hardware breakpoints. That is, the source debugger that drives the BDM/JTAG port sets a breakpoint by replacing the first byte or word of the instruction's opcode with a special instruction that places the chip in debug mode. This is much like ROM monitors that use an illegal opcode or similar instruction to invoke a breakpoint handler.

Most of the interfaces, though, also have a hardware breakpoint input pin. Drive this line high and the CPU halts execution of the firmware. Some vendors offer quite elaborate bus monitors (for those target systems that indeed have a viewable bus) that support complex break conditions ("break when routine `timer_isr` called after variable `foobar` written"). This is where ICE meets BDM, as quite a bit of ICE-like hardware is required.

So, the upside of a BDM or JTAG debugger boils down to this:

- A debugger on-board the chip eliminates all speed issues. It functions despite cache's complications. Even when the CPU is hidden in a huge ASIC, if just a few pins come out for the serial debugger, then designers will have some ability to troubleshoot their code.
- JTAG/BDM lets you set simple breakpoints, single-step, and examine and change memory and I/O . . . in short, everything you can do with a normal PC design environment, such as Microsoft's Visual C++.
- BDM-like solutions are a reasonable subset of a debugging methodology. They're so inexpensive that every developer can have the toolset. Some tool vendors properly promote these as nothing more than debugging adjuncts, devices designed for working on certain non-real-time sections of code. Their message is to "use the right tool for the right job—a BDM where it makes sense, and a full-function emulator for real-time troubleshooting."

Given that run control offers basic system access, breakpoints, and the like, what do we lose when we chose one of these over an ICE?

- Emulation RAM does not exist on BDMs. No serial debugger now extant or proposed offers any sort of memory that replaces your system ROM. To download code, you can relink so the code executes from your system RAM area, assuming there's plenty of free RAM space, or replace your ROM chips with RAM, which depending on your system design may or may not be possible. Another option is to mix tools, using a ROM emulator; download code to the emulator and test it via the BDM/JTAG port.

- Breakpoints, too, will not have the power and sophistication you may be used to with an ICE. Most such debuggers won't permit nested complex conditions, or pass counters, or even hardware (as opposed to software) breakpoints.
- Trace is probably the biggest loss when moving from an ICE to a serial debugger. Some tool companies have married logic analyzers to run control BDM/JTAG devices. The result is a trace-like output . . . but only in the cases where the CPU busses are available and probeable. However, a lot of work is now taking place to add limited trace capabilities to these products.

ROM Monitors

The oldest of embedded tools is still a viable and useful option for many projects. The ROM monitor is nothing more than a little bit of code that is linked into your target firmware. You allocate a communications port to the tool; it uses this port to interpret commands from the source debugger hosted on your PC.

The ROM monitor is generally a rather simple bit of code. It sends register and memory info to the PC and accepts downloaded code from the same source. Breakpoints are simple address-only types.

ROM monitors have the following wonderful attributes:

- They're cheap! The ROM monitor is a simple bit of code. Most of the cost of the debugger will be in the source-level debugger.
- The tool has no physical connection problems. Stick it in any system, no matter how fine the SMT pins or how deeply buried the CPU core lies.
- Speed problems just don't exist, since the monitor is just software running concurrently with the rest of your code.

The downsides to ROM monitors include:

- The tool requires exclusive access to a communications port; if a ROM monitor is in your future, be sure to add an extra comm port to the hardware just for the sake of the tool.
- The ROM monitor will consume other target resources such as ROM and RAM, and maybe some interrupts. In a big 32-bit system this is rarely a problem. If you're working in a 4k address space, these resources are usually too scarce to dedicate to the tool.
- There's always a setup/configuration problem, as you've got to link the tool into your code and connect it to your proprietary communications port.

- The ROM monitor will not work if the hardware is broken.
- Real-time instrumentation is weak. You just won't find trace or timing data in any ROM monitor product.

ROM Emulators

A significant problem with conventional emulators is that they are CPU-specific. Change from a 68332 to a 68340 and, even though the processor's architecture doesn't change, you'll need a new emulator—or at least a new multi-thousand-dollar pod. ROM emulators, instead, connect to your target system via a memory socket. They consist of a RAM array that mimics the ROM chip . . . while allowing you to download new code in a heartbeat. The serial port is built into the unit itself.

ROM emulators are so inexpensive that even when using some other debugging tool I keep a few around for those unexpected problems that always seem to surface.

ROM emulators continue to play an important role in embedded development for the following reasons:

- As ROM replacements they offer convenient overlay RAM. Especially in smaller systems, this may be critical so you can download code, rather than burn a dozen ROMs an hour.
- Most are very inexpensive—some go for just a few hundred dollars. This means every developer can have a reasonable debugging tool at hand.
- ROM emulators are processor-independent. The source debugger may change as you move from a 68000 to a 186, but the hardware element remains unchanged.
- Few, if any, target resources are required.

Problems include:

- Just as with an ICE, speed is an ever-increasing concern.
- The physical connection to the target system might be difficult if you're emulating SMT ROM devices. As with ICEs, many vendors do offer innovative connection strategies, but bear in mind that making a reliable connection may be difficult.
- The ROM socket does not provide any convenient way to set breakpoints! About half of the vendors do offer a breakpoint strategy; be sure the one you select won't leave you breakpoint-starved.

Oscilloscopes

Emulators, ROM monitors, and the like are great for viewing your code from the perspective of the CPU. Their tentacles into your target system stop at the CPU socket, so events occurring beyond that point (say, in an I/O device) are almost invisible. You can see the IN and OUT instructions and the transferred data, but it's pretty hard to check out timing relationships, or how the software interacts with the hardware.

Sure, most of these tools have external inputs that you can couple to any point in the system. Few programmers use them. Perhaps this is because the display is so static. You have to actively recollect data and then tediously sort it all out. For example, if you feed an external input to a real-time trace buffer, you'll collect tons of bus activity that may or may not be important.

If all you really care about is the relationship between two events (say, a switch closure and the resultant interrupt), why dig through thousands of cycles? It is important to arm ourselves with as many tools as possible. No one tool is perfect for every problem.

One of my all-time favorite software debugging tools is the oscilloscope, colloquially known as the "scope." Hardware guys seem to have a scope attached as a pseudopod to one arm. Any development lab is invariably filled with benches of scope-happy troubleshooters probing the mysteries of some electronic marvel. The software community seems less comfortable with this tool, which is a shame because it can painlessly yield crucial information about the operation of your code.

A scope is really nothing more than a device that displays one or more signals. Most can simultaneously show two independent values.

The scope's raison d'être is displaying the signals' voltage (amplitude) over time.

A simple time-varying signal is the power coming from your wall outlet. This is a 60-Hz sine wave (i.e., the voltage smoothly rises from 0 to 120 and back to zero again 60 times a second). It moves too fast to follow with a voltmeter. On a scope display, the waveform's voltage at any point in time is crystal clear.

Software folks used to working with only a keyboard are sometimes intimidated by the sea of knobs on any decent scope's front panel. A bit of experience makes working with this tool natural.

From the user's standpoint the average scope has three major sections. A "vertical" amplifier sets the display's up/down limits. The "horizontal" portion controls the beam's left/right scanning. "Trigger" circuitry synchronizes the scan to your input waveform.

Given that the scope is a general-purpose tool used by RF engineers, digital computer designers, and even software gurus, it has to accept a wide range of inputs. Computer people work mostly with 5-volt levels (i.e., a zero is about 0 volts; a one is 3 to 5 volts). Audio engineers might need to measure millivolt levels. Your embedded system probably detects or generates some sort of real-world data, which is probably not in the 0- to 5-volt scale.

Thus, the scope's Vertical section is born. The run-of-the-mill two-channel scope has two identical vertical sections.

A BNC connector (like the kind used in thin Ethernet applications) connects to the scope probe. The signal sensed by the probe runs to the vertical amplifier, which increases the input from perhaps a few volts to several hundred, which is ultimately applied to the plates in the CRT.

Like any good amplifier, each vertical channel has an amplitude control (i.e., the same thing as a volume control in your stereo). Unlike a volume control, it has an exact calibration associated with each position. Set the knob to, say, 2 volts/division, and a 4-volt signal will move the beam up two divisions. Divisions are denoted by a grid of boxes on the CRT so you can easily measure levels.

Each channel has a "position" control that lets you move the rest position of the beam up or down to the most convenient point. If you wanted to measure voltage, with no signal applied, set the beam right on one of the division marks on the screen. Then, count how many boxes the waveform occupies. Convert divisions to voltage using the setting of the amplitude control.

The position control lets you move the beam all the way off the screen. It can be pretty challenging to find the damn beam at times, so a "beam find" button brings it into view, giving you an idea which way to move the position controls.

A channel selector lets you put either channel 1 or channel 2 on the screen. Most software work involves measuring the relationship between two inputs, so you'll select "both." Two sweeps will pop up. Use the two sets of amplitude and position knobs to control each channel independently.

Controlling up and down beam deflection is only half of the problem. The Horizontal Amplifier sweeps the dot back and forth across the screen. Note that you only see the left-to-right deflection; the return sweep is very fast and is never displayed.

In software debugging I hardly ever care about amplitude, since mostly I'm looking for the input's shape or duration. If the amplitude is wrong, generally there is a hardware problem. I set up the vertical controls just to get a decent-sized waveform and then mostly ignore them.

Timing, though, is always crucial. The horizontal system doesn't just randomly move the beam back and forth; it does so in a highly regular and measurable manner.

Generally the biggest knob on a scope is the one labeled something like "Time/Division." Try cranking it through all of its positions. Go all the way counterclockwise: the beam will be a single dot, either stopped or moving very slowly to the right.

As with the amplitude control, this switch is calibrated. The slowest sweep rates (all the way counterclockwise) might be as much as 5 seconds per division. Slowly rotate the knob and watch as the dot picks up speed. 5 sec/div, 2 sec/div, 1, .5, .2, .1—pretty soon the dot will be moving so fast it will start to look like a line. Rotate it all the way. Now, the dot is moving at perhaps 50 nanoseconds per division. That's fast!

The horizontal system is frequently called the "time base," because it provides all basic timing functions to the scope.

A cardiac monitor is nothing more than a specialized oscilloscope. A very slowly moving beam shows the patient's heart rate. The signal beats only 70 times/sec, so a slow rate is best to represent the input.

Suppose the signal moves not at 70 beats/sec, but at 7 million (say, for a hummingbird on speed). At the slow sweep rate of the cardiac monitor the beam will move up and down so fast compared to the left-to-right sweep that a band of light will appear. You'll see no recognizable signal. Crank up the sweep rate. The band will eventually resolve itself into the familiar cardiological shape. At first, the signal will be all squished together. Perhaps three beats will be in each division. Rotate the knob again. Now, only one beat is in a division. With each rotation the horizontal image expands. With each rotation you can still measure the beat frequency by counting divisions and applying the Time/Division parameter listed on the control.

The Horizontal control, then, lets you pick a sweep rate that generates a recognizable picture of the signal you are measuring.

There's always one little detail to complicate matters. So far we've ignored the issue of synchronizing the sweep to the signal.

In the case of the cardiac input, suppose on one sweep the beam starts off on the left side of the screen when the signal is halfway up the slope, and the next sweep starts when the input is at 0 volts. The position of the display will shift left or right on every sweep, creating an image impossible to focus on.

Unless the sweep starts at the same point on the input signal each time, the display will look like a meaningless jumble. In the bad old days before trigger circuits, people tried to tune the sweep frequency to exactly match the input, but this is hard to do at best, and is pretty much impossible with digital circuits.

The modern solution is the third component of any decent scope. The "Trigger" controls let you pick the sweep starting point.

Generally, selector switches let you pick AC or DC coupling, trigger level, holdoff, slope, and trigger source selection. The correct procedure is to select a reasonable source (channel 1 or 2: which one do you want to use to start the sweep?), and then start twiddling knobs until the display stabilizes.

Sure, it makes sense to follow some semblance of a procedure. Select a (+) slope if you want to see the upgoing edge of the input at the very left side of the screen. Select (−) slope to position the downgoing edge there.

Start twiddling with the holdoff control set to OFF (usually all the way counterclockwise). Most of the magic will be in the Trigger knob, which requires a delicacy of touch that takes some practice to develop.

Triggering on any repetitive signal is pretty easy, because the differences from sweep to sweep are small. Digital signals are more challenging. A constantly changing pulse stream is all but impossible to capture on a scope.

Scoping Tricks

One of the worst mistakes we make is neglecting probes. Crummy probes will turn that wonderful 1-GHz instrument into junk. Managers hate to spend a lot on probes when they see them drooling onto the floor, mixed with all of the other debris. Worse, we always immediately lose the tips and other accessories acquired at great expense, and so connect to a node using a 12-inch clip lead hastily purchased at Radio Shack.

Then, after destroying a couple of chips by accidentally shorting things to ground with that nice alligator ground clip mounted on the probe, we tear it off in frustration, losing it as well. Tip: If you really don't intend to use the ground connection, clip that alligator lead to itself, keeping it out of harm's way but instantly available for use.

Take care of your probes. Keep them off the floor; don't let your chair roll over the leads, squishing the coax and changing its impedance. Buy decent ones *before* every probe in the shop falls apart. After trying all of the cheap varieties found in general electronic catalogs, I now swallow hard and spend the $150 needed to get high-quality probes from Tektronix or HP.

Here's another tip: When you're using a scope, if a signal looks weird, maybe there's something wrong! Avoid the temptation to rationalize the problem. Instead of blaming the signal on a lousy ground, quickly connect that ground clip and test your assumption.

Never accept something that looks awful. Either convince yourself that it's actually OK, or find the source of the problem.

Walk through your lab. You'll find that most of the digital folks have their vertical amplifiers set to 2 volts/division, which eases displaying two traces simultaneously. Unfortunately, too many of us seem to think the vertical gain knob is welded into position. It's hard to distinguish a valid zero from one drooling just a little too high with so little resolution. Flip to 1 V/division occasionally to make sure that zero is legitimate.

Every instrument is a lying beast, a source of both information and disinformation. The scope is no exception. A 100-MHz scope will show even a perfect 50-MHz clock as a sine wave, not in its true square form. Digital scopes exhibiting aliasing sweep too slowly (below the Nyquist limit) for a given signal, and that 50-MHz clock may look like a perfect 1-kHz signal, causing the inexperienced engineer to go crazy searching for a problem that just does not exist. Try this experiment: measure a 10- or 20-MHz clock on a digital scope. Crank the sweep rate slower and slower. You'll inevitably reach a point where the scope shows a near-perfect square wave several orders of magnitudes slower than the actual clock frequency. This is an example of aliasing, where the scope's sampling rate yields an altogether incorrect display. I'm sure many folks have heard a claim such as, "This 16-MHz oscillator is running at 16 kHz! Can you believe it?" Don't. Check your settings first.

We digital folks deal in ones and zeroes . . . and tristates. Each condition means something. When troubleshooting, you've got to know which of these three (not two) states a node is in. Our best tool is the scope, yet it is inherently incapable of distinguishing the tristate condition.

In the good old days of LS technology you could be pretty sure a tristated signal would show up at around 1.5 volts—somewhere between a zero and a one. With CMOS this assurance is gone, yet most engineers blithely continue to assume that zero volts means zero. It just ain't so.

My solution is a little tool I made: a 1k resistor with a clip lead on each end. Mine is nicely soldered together and covered with insulation to avoid shorts. To tell the difference between a legal state and high impedance, clip the tool to the node and alternately touch the other end to Vcc and then ground. If the node moves more than a trifle, something is wrong. The scope, plus my tool, lets me identify all three possible states. Without the tool I'm guessing, and guessing while troubleshooting always sends you down time-consuming blind alleys.

You can use a variation of this approach when troubleshooting an intermittent problem. If the silly thing refuses to fail when you're working on it—a sure bet, given the perversity of nature—run your fingers over the board's pins. A purely digital board should continue to run despite the slight impedance changes brought about by your fingers, yet these may be enough to drive a floating pin to the other state, possibly creating the failure you are looking for.

On SMT boards it's tough to get at a device's pins. If there's one pin you are suspicious of, touch it with an X-Acto knife. The sharp blade will precisely align with any tiny pin, and its metal handle will conduct your body impedance to the node. Sometimes I'll connect my trusty pull-up/pull-down clip lead to the knife itself to exercise the node more deterministically.

No scope will give decent readings on high-speed digital data unless it is properly grounded. I can't count the times technicians have pointed out a clock improperly biased 2 volts above ground, convinced they found the fault in a particular system, only to be bemused and embarrassed when a good scope ground showed the signal in its correct 0- to 5-volt glory.

Yet most scope probes come with crummy little ground lead alligator clips that are impossible to connect to an IC. Designers all too often insert a clip lead in series just to get a decent "grabber" end. Those extra 6 to 12 inches of ground lead will corrupt your display, sometimes to such an extent that the waveform is illegible. Cut the alligator clip off the probe and solder a micro grabber on in its place.

Ask an experienced scoper to work with you for a couple of hours. Have the mentor randomly shuffle the controls; then try to bring the display back and stabilize it. Try probing around a battery-operated radio (where there are no dangerous voltage levels!). Look at signals. Fiddle with the trigger controls and time base to stabilize and examine them.

Fancy Tools, Big Bucks?

As an ex–tool vendor I can't count the times I've heard, "Well, we really need decent equipment, but my boss won't let me spend the money."

It matters little what equipment we're talking about. Once I wrote an offhand comment about companies who won't upgrade computers. An avalanche of email filled my electronic in-box, from developers saddled with 386-class machines in the Pentium age. We live in front of our computers, spending hours per day with them. It's incomprehensible to me that a business won't provide very expensive engineers new machines every two years. I've seen compile times shrink from tens of minutes to tens of seconds when transitioning just one generation of computers; surely this translates immediately into real payroll savings and faster development times!

Yes, we have an insatiable appetite for new goodies. Glittering new scopes, emulators, logic analyzers, and software tools fill our thoughts much as kids dream of Tonkas and Barbies. Very often, though, the gap between what we want and what we get is as wide as the Grand Canyon.

Now, I know the cost and scarcity of capital. Just try going to the bank, hat humbly in hand, looking for working capital when you really need it. Venture capital is the seed of high tech, but is much less available than people realize.

There's never enough money, especially in smaller businesses, so every decision is a financial tradeoff between competing needs.

I also know the cost of payroll. It's by far the biggest expense in most technology businesses. Yet many managers view payroll as a sunk cost. Years ago my boss told me, "I have to pay you anyway, but to buy that scope costs me real money."

Well, no, actually, he didn't have to pay me or any of the engineers. He had options: do less engineering with fewer people and save on salary. Use us inefficiently and ignore the costs. Work to improve our efficiency and either get products out faster or get the same work done with fewer people.

This concept of payroll as a fixed cost is a myth, one that destroys too many technology companies. Managers do have the ability to manage this cost, the biggest one of all, effectively. It's not easy and it's never "done"; effective management requires an intimate understanding of the processes involved, a willingness to experiment and tune, and a dedication to a never-ending quest to find lots of 1 and 2% improvements, as the magic 20% efficiency improvements are indeed rare.

Our culture of absorbing payroll as a fixed expense means we battle for weeks over $10,000 tool costs while ignoring, or accepting, $1 million in salary costs.

Perhaps this is symptomatic of uninformed managers and exhibits itself in every area of development. One friend who makes a living designing products as a contractor tells me story after story of companies that happily spend a quarter million dollars on tooling for the product's plastic box, yet balk at a quote for $30k in custom firmware.

I see an increasing number of companies embracing the noble ideal of "doing more with less" without understanding that sometimes spending a bit on tools is the fastest route to that ideal.

You can't pick up a trade magazine today without seeing the industry's mantra—Time To Market—gracing every article and ad. All sorts of studies indicate that getting a product out first is the best way to gain market share and profitability. Whether this is true or not makes little difference; the important point is that management has universally bought into the concept, leaving it up to engineering to somehow "make it so."

The time-to-market furor explains surveys that show development time to be the number one priority of many engineering departments, with cost usually running third after quality. Whether we agree with the goals or not, it is at least a reasonable ranking of priorities.

Get it done fast. Do a good job. And then worry about costs. These are the constraints we're working under, in order.

But we can't develop a realistic plan without considering all of the facts. One is that salaries continue to rise, especially now, and especially for highly trained and scarce engineers. None of us can control this.

Fast, gotta be fast. Cheap, too—somehow we have to save bucks wherever we can. OK . . . now what?

Astonishingly, more and more companies are making decisions like: no tools. Poor tools. Or, let's pick a chip that has no tools, or for which decent tools are a but a dream.

How on earth are we supposed to be fast with inadequate tools? Won't costs skyrocket as we spend more time struggling to find bugs— bugs that are more evasive than ever as products get more complex—using what amounts to toys?

In the face of increasing salaries, more complex products, and terrifying schedules, all too often the question *"How* are we going to get the work done?" never gets answered honestly.

Yet, as you read this today, hundreds of companies pursue development strategies that are doomed to cost too much and take too long. Some use custom microprocessors—for good reasons and bad—and build their own compilers and debuggers. I'm not saying this is necessarily wrong; it's just costly. Some of these businesses understand and manage the issues; others just yell louder at the developers to meet the schedule.

I've seen months spent gluing CPUs inaccessibly into the core of a monster ASIC, without the least thought given to debugging . . . and then the hardware guys present the firmware folks with this fait accompli and only two months left in the schedule.

We *must* look at the technology challenges posed by the parts we choose, and then at our options for building the system and then finding bugs. We *must* find or invent ways of achieving our fast–quality–cheap goals before committing to a difficult or impossible technology.

And, management must understand that time costs money—real money, not just sunk costs. Further, crummy development environments never yield faster product introductions.

This is not a Dilbert-like rant against managers. We're all infatuated with the latest technology, and we all are convinced that, this time, bugs won't be as big of a problem as last time.

Embedded processors will continue to get faster and more highly integrated—and will generally become much tougher to work on than those of yesteryear. That's a fact as sure as salary inflation and time-to-market pressures.

It's largely up to the developers doing the work to educate management, and to make intelligent decisions yielding debuggable products.

Often we are perceived as wanting everything without decent justifications. Faster computers, private offices, better software tools. Without educating our bosses about how these things save them money, we'll lose most battles.

A common joke is the "capital equipment justification," all too often more an exercise in creative writing than in fact gathering and analysis. Sometimes tool vendors will present you with spreadsheets of savings from using their latest widget, but none of us really trusts these figures. It's far better to use hard-hitting, quantitative data accumulated from your own hard-won experience. Don't have any? Shame on you!

One well-known bug reducer is recording each bug, stopping and thinking for a few seconds about how you could have avoided making the mistake in the first place. Take this a step further and think through (and record!) how you found it, using what tools. Log it all in an engineering notebook as you work; it's a matter of a few seconds' time, yet will help you improve the way you work. This notebook will also serve as the raw data for your cost justifications. If that cruddy freeware compiler generated a bad opcode that took a day to find, a little math quickly will show how much money a multi-thousand-dollar commercial package would save.

As you educate management, educate yourself, and remember those lessons when you're the boss!

Years ago I worked for a small, 100-person outfit that experienced a wealth of financial difficulties. Half of the phone calls were from angry creditors. The bank was perpetually on the brink of closing us down. Still, our small engineering group always had a reasonable set of tools. Good scopes then cost upwards of $10,000, a lot of money in 1975 dollars. We even managed to get one of Intel's first microprocessor development systems. Though we engineers had to cajole and plead with management for the tools, we did get them, and developed an expectation that we'd always have access to whatever the job needed.

Then I started consulting.

Suddenly, those wonderful tools we had so long taken for granted were no long available. My partner and I shared an old Tektronix 545 scope (that used vacuum tubes—you know, those glass-shelled things with filaments and high voltages). We scraped up enough money to build an emulator—such as it was—from mail-ordered Multibus boards. A $400 CRT terminal and daisy-wheel printer were all we could afford in the way of new capital equipment.

We learned all sorts of ways to extract information from systems, pouring loads of time into projects instead of cash.

Then I met a fellow whose high-school kid had a lab of sorts in his home. He had a new Tektronix scope! I was flabbergasted. Though the unit wasn't top-of-the-line, it sure beat the antique I was saddled with.

A few discreet questions turned up the fact that he rented the scope, for a lousy $50 a month. Somehow it had never occurred to me that there were options other than coming up with thousands in cash. This kid had shown me that the quest to obtain the right tools is a *problem*, one like any other problem we run into in engineering and life, one that takes a bit of creative energy to solve.

Ain't America grand? Easy credit, available to practically any warm body, means we can satisfy practically any whim . . . as far too many of us do until the inevitable day of reckoning comes.

Look at the computers advertised in any PC magazine. Every ad has a caption giving the low, low monthly payment they'll require. If your business has any income at all, then the hundred a month or so for a high-end machine is a pittance.

Test equipment vendors all offer similar plans. You'd be surprised how low the monthly payments on a scope are, when spread over three to five years.

Most companies will bend over backwards to finance your purchase. Those that have no in-house financing ability work with third-party financial outfits. Test equipment companies really want you to have their latest widget, and they'll do practically anything to help you purchase it.

Renting is a traditional means to get access to equipment for short periods of time. However, unless you're quite convinced that the project will end as planned, be wary of rentals. Few short-term projects fail to increase in scope and duration. Since rentals generally cost around 10% of the unit's purchase price per month, once the project slips more than a quarter, you may have been better off buying than renting.

Leases are the most attractive way to get equipment you can't afford to buy outright. A lease with buyout clause is nothing more than a financed purchase. It may have certain tax benefits as well, though this part of the law changes constantly.

Even for a single scope you can get leases amortized over practically any amount of time. Three years is a common period. The monthly payment will be something like 3% of the unit's purchase price per month. A $5000 logic analyzer will set you back around $200 per month. For less than your car payment you can get a nice scope and logic analyzer. Unlike the car, neither will wear out before the payments are up.

Sometimes it makes sense just to purchase gear outright, especially since the IRS permits you to expense $17,500 of capital equipment per year. When cash is tight, consider getting used, refurbished test equipment. A number of outfits sell reconditioned gear for around 50 cents on the dollar. Good test equipment lasts almost forever.

One acquaintance has just a shell of a company, a so-called "virtual corporation" that changes dynamically as business ebbs and flows. He shares an office suite with other like-structured organizations. All are in the digital business and use a common lab area with shared test equipment. For small outfits, this is a neat way to make the dollar go a lot further.

Tool Woes

After reading the glossy brochures and hearing the promises of suited tool salespeople, you're no doubt convinced that their latest widget will solve all of your debugging problems in a flash.

Not.

Be wary of putting too much faith in the power of tools. Too many engineers, burned by previous projects, do a good job of surveying the tool market and selecting a reasonable development environment, but then put all their hopes of debugging salvation in the toolchain.

The fact is, vendors tend to overpromise and underdeliver. Perhaps not maliciously, but their advertisements do play into our desperate searches for solutions. The embedded tool business is a very fragmented market. With hundreds of extant microprocessors, the truth is that typically only dozens to (maybe) a couple of thousand users exist for any single tool. With such a small user base, bugs and problems are de rigueur.

I write this as an ex–tool vendor who strongly believes that an important component of productivity comes from using a first-class development environment. But, as an ex-vendor, all too often I saw engineers who expected that spending five or ten thousand on the gadget would miraculously solve most problems. It just ain't so. Buy the right tools, but understand their inherent limitations.

Overcome limitations with clever designs, using a deep understanding of where the problems come from. Here's a collection of ideas drawn from bitter experience:

Reliable Connections

In the good old days microprocessors came in only a few packages. DIP, PGA, or PLCC, these parts were designed for through-hole PC boards with the expectation that, at least for prototyping, designers would socket the processor. Isolating or removing the part for software development required nothing more than the industry-standard chip puller (a bent paper clip or small screwdriver).

Now tiny PQFP and TQFP packages essentially cannot be removed for the convenience of the software group. Once you reflow a 100-pin device onto the board, it's essentially there forever.

Part of the drive toward TQFP is the increasing die complexity. That tiny device is far more than a microprocessor; it's a pretty big chunk of your system. The CPU core is surrounded with a sea of peripherals—and sometimes even memory. Replace the device with a development system, and the tool will have to replace both the core and all of those high-integration devices.

Take heart! Most semiconductor vendors are aware of the problem and take great pains to provide work-arounds.

There's no cheap cure for the purely mechanical problem of connecting a tool to those whisker-thin pins, but at least the industry's connector folks sell clips that snap right over the soldered-on processor. The clip translates those SMT leads to a PC board with a PGA or header array that your tools can plug into. Before starting any design, get a copy of Emulation Technology's catalog. Though their products are horrifically expensive, they offer a very wide range of adapters and connection strategies.

Another good source for connection ideas is the logic analyzer arena. Both HP and Tektronix are starting to standardize their analyzer cables on AMP's "Mictor" connector, a very small, very high-density, controlled impedance device. If you surround your CPU with Mictors (being careful to match the pinouts used by the analyzer vendors), then probing becomes trivial: just plug the analyzer cables in directly. If you're frustrated with logic analysis because of the agony of connecting 50 or 100 little clip leads (half of which pop off at inconvenient times), take heart, as the Mictor goes directly into the main analyzer cables, bypassing the clips altogether.

A Canadian company had a PCMCIA-based product whose CPU's whisker-thin TQFP leads defeated every ICE connection attempt. Their wonderfully clever solution was to design the card with a large extra connector—a 100-pin header—to which all of the CPU signals went. This, of course, doubled the size of the board. The connector sat at the far side of the board, outside of the PCMCIA's nominal form factor (i.e., when the board was plugged into a laptop computer, the connector protruded into space outside of the PC). The engineers ensured that the connector's pinout exactly matched that of the emulator they selected, so the ICE's pod plugged in with no adaptors or other reliability reducers. When it came time to ship the product they cut the connector off, and the board down to size, with a bandsaw. Production versions, of course, were proper-sized cards without the connector.

If your product uses a card cage, no doubt the board-to-board spacing is insanely tight. Too often extender cards don't work, since the CPU becomes unstable driving the extra long lines. Just debugging the hardware is hard enough—try slipping a scope probe in between boards! It's not unusual to see a card with a dozen wires hastily soldered on, snaked out to where the scope or logic analyzer can connect.

Why make life so hard? Either design a robust processor board that works properly on an extender, or come up with a mechanical strategy that lets you put the CPU near the end of the cage, with the cage's metal covers removed, so you and the software people can gain the access so essential to high-productivity debugging.

One DOD system's card cage is so tightly packed into the rack of equipment that the developers could only remove the "wrong" (i.e., circuit) side of the card cage cover. Their solution: solder the processor socket on the circuit side of the board, and then make a pin swapping jig for the logic analyzer. Using a ROM emulator in a similarly tight situation? Consider the same trick, inverting one or more ROM sockets.

Make sure the CPU (when using an ICE or logic analyzer) or ROM sockets (ROM emulator) are positioned so it's possible to connect the tool. Be sure the chip's orientation matches that needed by the emulator or analyzer.

Nonintrusive Myths

Debugging tool vendors all promote the myth of "nonintrusive tools." In fact, we demand just the opposite—what could be more intrusive, after all, than hitting a breakpoint?

Other forms of intrusion are less desirable but inevitable as the hardware pushes the envelope of physical possibilities. If you don't recognize these realities and deal with them early, your system will be virtually undebuggable.

Don't push the timing margins. All emulators eat nanoseconds. With no margin the tool will just not work reliably. I've seen quite a few designs that consume every bit of the read cycle. Some designers convince themselves that this is fine—the timing specs are worst-case scenarios met at max or min temperatures, leaving a bit of wiggle room for the tool. As speeds increase, though, IC vendors leave ever less slop in their specifications. It's dangerous to rely on a hope and a prayer.

Before designing hardware, talk to the tool vendor to learn how much margin to assign to the debugger. Typically it makes sense to leave around 5 nsec available in read and write cycle timing. Wait states are another constant source of emulator issues, so give the tool a break and ease off on the times by four or five nanoseconds there, as well.

Fact: if you don't leave sufficient margin, the system will be virtually undebuggable. Now, BDMs and ROM monitors will generally work in marginless designs, but you'll give up the ability to bring up dead hardware and track real-time firmware flow.

Be wary of pull-up resistors. CMOS's infinite input impedance lures us into using lots of ohms for the pull-ups. Remember, though, that when you connect any sort of tool to the system, you'll change the signal loading. Perhaps the tool uses a pull-down to bias unused inputs to a safe value, or the signal might go to more than one gate, or to a buffer with wildly different characteristics than used on your design. I prefer to keep pull-ups to 10k or less so the system will run the same with and without an emulator installed.

If you use pull-down resistors (perhaps to bias an unused node such as an interrupt input to zero, while allowing automatic test equipment to properly bias the node in production test), remember that the tool may indeed have a weak pull-up associated with that signal. Use too high of a resistance and the tool's internal pull-up may overcome your pull-down. I never exceed 220 ohms on pull-downs.

Synchronous memory circuits defeat some emulators. These designs ignore the processor's read and write outputs, instead deriving these critical signals from status outputs and the clock phase. Vadem, for example, makes chip sets based on NEC's V30 whose synchronous timing is famously difficult for ICEs.

This sort of timing creates a dilemma for ICE vendors. What sorts of signals should the emulator drive when the unit is stopped at a breakpoint? A logical choice is to drive nothing: put read, write, and all other control signals to an idle, nonactive state. This confuses the state machine used in the synchronous timing circuits, though; generally the state machine will not recover properly when emulation resumes, and thus generates incorrect reads and writes.

Most emulators cannot afford to completely idle the bus, anyway, as it's important to echo DMA and refresh cycles to the target system at all times.

Since the processor in the ICE usually runs a little control program when sitting still at a breakpoint, another option is to echo these read/write cycles to the bus. That keeps the state machine alive, but destroys the integrity of the user's system because internal emulator write cycles trash user memory and I/O.

Another possibility is to echo the cycles, but fake out write cycles. When the emulator's CPU issues a write, the ICE drives an artificial read to the target. Unhappily, on many chips read and write cycles have somewhat different timing, which may confuse the user's state machine.

None of these solutions will work on all CPUs and in all user systems. If you really feel compelled to use a synchronous memory design, talk to the emulator vendor and see how they handle cycle echoing at a breakpoint.

Consider adding an extra input to your state machine that the emulator can drive with its "stopped" signal and that shuts down memory reads and writes. Talk timing details with the vendor to ensure that their "stopped" output comes in time to gate off your logic.

Add Debugging Resources

Debugging always steals too much time from the schedule. This fact implies that we've got to anticipate problems when designing the hardware, and take every action possible to ease troubleshooting.

Always—unless your system is so cost constrained that a buck is a huge deal—add an extra output port to the system, one dedicated just to debugging. Why?

- As we saw in Chapter 4, a very effective and inexpensive way to measure system performance is to instrument your code. Add a line that sets a bit—on this I/O port—high when in an ISR to measure ISR time. Diddle another I/O bit in the idle loop to measure overall system loading.
- Toggle one of the bits when the system resets. As I said in Chapter 6, a watchdog time-out is a serious event. If your system auto-

matically recovers from the watchdog reset, you surely need some way, during debug, to see that the time-out occurred.
- When your tools are not working well, or perhaps you've simply lost faith in them, you can still track overall program flow by assigning an 8-bit number to each important function. Output this number to the debug port when the function starts. Collect the data in the logic analyzer and you'll instantly see what executes when, and for how long.
- Connect one or more of the more I/O bits to LEDs, and instrument the code to signal system state. Most tools do a poor job of reading out state; generally you'll have to stop the code or something similar. The LED bank instantly shows things like, "It's doing WHAT???!!!!!"

If your main debug strategy revolves around a full-blown emulator, if at all possible go ahead and add the BDM or JTAG connector (if the CPU supports it). The cost is vanishingly small, and the option of doing BDM debugging when the ICE falls flat or fails may save a lot of money and time.

Conversely, if a BDM will be the main tool, add a connector (like the Mictor) so that you can connect a logic analyzer for tracking real-time events. It's so terribly difficult to use analyzers via their standard multitude of clips that we leave it as a last resort; if it's easy to connect, we'll use the tool at the appropriate times.

ROM Burnout

Remember that every tool affects system operation in some manner. Never wait until the night before shipping to test the system from ROM. Make burning a ROM or loading the Flash a regular part of the test procedure.

- Debugging tools invariably have a different size of emulation RAM than your target system's ROM space (this is true using an ICE or a ROM emulator, or even if you relink your code to run from your system RAM area). If the code grows to exceed target ROM space, it may run just fine from the (probably bigger) emulation RAM area.
- The compiler's runtime package or constants might be improperly initialized. Many C compilers require a startup procedure that copies some critical variables to RAM. When you're debugging, you'll generally replace system ROM with RAM merely to support

quick code downloads. If the initialize is not correct, since you're debugging from RAM things may work just fine . . . until that first ROM burn.
- Often hardware problems mean that the ROM sockets on your target just don't function properly. This may be due to wiring or design problems . . . or even to buggy code. An improperly configured chip select signal, for example, may not create any problems working from emulation RAM, but will crash the code after the ROM burn.

Be wary of the converse situation: the code runs fine from ROM but not from emulation RAM. All too often a wandering pointer causes erratic writes over ROM space, surely a very bad thing. This happens so often that we should take a defensive posture and regularly look for such problems. Depending on your tools, this is pretty trivial:

- Many emulators support modes that will automatically watch for writes to code space. If the tool doesn't explicitly include such a resource, you can still usually configure one of the complex breakpoints to break on any "write to address between X and Y," where X and Y represent the range of addresses of code.
- Occasionally checksum your code. That is, download the code and compute a checksum of the image using the tool's checksum command. Run the application for a while and recompute the checksum. Any change generally indicates a serious problem.
- Wandering pointers are such a common problem, and are so difficult to find, that there's a lot to be said for leaving a logic analyzer connected that's configured to watch for errant memory accesses. The wonderful triggering capability of these tools means it's easy to set up multiple conditions that watch for any stupid memory access. What do I mean by "stupid"? A write to code space. A fetch from data areas. Any access to unused memory. Trigger on these three conditions and you'll catch a huge percentage of wandering pointers.

CHAPTER **8**

Troubleshooting

There comes a time in any project when your new design, both hardware and software, is finally assembled, awaiting your special expertise to "make it work." Sometimes it seems like the design end of this business is the easy part; troubleshooting and debugging can make even the toughest engineer a Maalox addict.

You can't fix any embedded system without the right world view: a zeitgeist of suspicion tempered by trust in the laws of physics, curiosity dulled only by the determination to stay focused on a single problem, and a zealot's regard for the scientific method.

Perhaps these are successful characteristics of all who pursue the truth. In a world where we are surrounded by complexity, where we deal daily with equipment and systems only half-understood, it seems wise to follow understanding by an iterative loop of focus, hypothesis, and experiment.

Too many engineers fall in love with their creations only to be continually blindsided by the design's faults. They are quick to overtly or subconsciously assume that the problem is due to the software (and vice versa), the lousy chips, or the power company, when simple experience teaches us that any new design is rife with bugs.

Assume it's broken. Never figure anything is working right until proven by repeated experiment; even then, continue to view the "fact" that it seems to work with suspicion. Bugs are not bad; they're merely a test of your troubleshooting ability.

Armed with a healthy skeptical attitude, the basic philosophy of debugging any system is to follow these steps:

```
For (i=0; i< # findable bugs; i++)
 {
 while (bug(i))
  {
    Observe the behavior to find the apparent bug;
    Observe collateral behavior to gain as much
    information as possible about the bug;
    Round up the usual suspects;
    Generate a hypothesis;
    Generate an experiment to test the hypothesis;
    Fix the bug;
  };
 };
```

Now you're ready to start troubleshooting, right? Wrong! Stop a minute and make sure you have good access to the system. No matter how minor the problem seems to be, troubleshooting is like a bog we all get trapped in for far too long. Take a minute to ease your access to the system.

Do you have extender cards if they're needed to scope any point on the board(s)? How about special long cables to reach the boards once they are extended?

If there's no convenient point to *reliably* clip on the scope's ground lead, solder a resistor lead onto the board so you're not fumbling with leads that keep popping off.

Some systems have signals that regulate major operating modes. Solder a resistor lead on these points as well, as you'll surely be scoping them at some point. This small investment in time up front will pay off in spades later.

Use the advice in the last chapter to ensure that your software is as probeable as the hardware.

Let's cover each step of the troubleshooting sequence in detail.

Step 1: *Observe the behavior to find the apparent bug.*

In other words, determine the bug's symptoms. Remember always that many problems are subtle and exhibit themselves via a confusing set of symptoms. The fact that the first digit of the LCD fails to display may *not* be a useful symptom—but the fact that none of the digits work may mean a lot.

Step 2: *Observe collateral behavior to gain as much information as possible about the bug.*

Does the LCD's problem correlate to a relay clicking in? Try to avoid studying a bug in isolation, but at the same time be wary of trying to fix too many bugs at the same time. When ROM accesses are unreliable and the front panel display is not bright enough, address one of these problems at a time. No one is smart enough to deal with multiple bugs all at once—unless they are all manifestations of something more fundamental.

Step 3: *Round up the usual suspects.*

Lots of computer problems stem from the same few sources. Clocks must be stable and must meet very specific timing and electrical specs . . . or all bets are off. Reset too often has unusual timing parameters. When things are just "weird," take a minute to scope all critical inputs to the microprocessor, such as clock, HOLD, READY, and RESET.

Never, never, never forget to check Vcc. Time and time again I've seen systems that don't run right because the 5-volt supply is really only putting out 4.5, or 5.6, or 5 volts with lots of ripple. The systems come in after their designers spent weeks sweating over some obscure problem that in fact never existed, but was simply the ghostly incarnation of the more profound power-supply issue.

Step 4: *Generate a hypothesis.*

"Shotgunners" are those poor fools who address problems by simply changing things—ICs, designs, PAL equations—without having a rationale for the changes. Shotgunning is for amateurs. It has no place in a professional engineering lab. And, as noted in Chapter 2, the software equivalent of shotgunning is making changes without a deep understanding of the bug. Use an engineering notebook to break the vicious "change/test" cycle.

Before changing things, formulate a hypothesis about the cause of the bug. You probably don't have the information to do this without gathering more data. Use a scope, emulator, or logic analyzer to see exactly what's going on; compare that to what you think *should* happen. Generate a theory about the cause of the bug from the difference in these.

Sometimes you'll have no clue what the problem might be. Checking the logical places might not generate much information. Or, a grand failure such as an inability to boot is so systemic that it's hard to tell where to start looking. Sometimes, when the pangs of desperation set in, it's worthwhile to scope around the board practically at random. You might find a floating line, an unconnected ground pin, or something unexpected. Scope around, but always be on the prowl for a working hypothesis.

Step 5: *Generate an experiment to test the hypothesis.*

Construct an experiment to prove or disprove your hypothesis. Most of the time this gets resolved in the process of gathering data to come up with the theory in the first place. For example, if the emulator reads all ones from a programmed ROM, a reasonable hypothesis is that CS or OE is not toggling. Scoping the pins will prove this one way or the other, though now you'll need another hypothesis and experiment to figure out why the selects are not where you expect to see them.

Sometimes, though, the hypothesis–experiment model should be much less casually applied. When Intel started shipping the XL version of the 186 (supposedly compatible with the older series), I had a system that just would not start with this version of the CPU. Scoping around showed the processor to be stuck in a weird tristate, though all of its inputs seemed reasonable. One hypothesis was that the 186XL was not coming out of reset properly, an awfully hard thing to capture since reset is a basically non-scopable one-time event. We finally built a system to reset the processor repeatedly, to give us something to scope. The experiment proved the hypothesis, and a fix was easy to design.

Note that an alternative would have been to glue in a new reset circuit from the start to see if the problem went away. Problems that mysteriously go away tend to mysteriously come back; unless you can prove that the change *really* fixed the problem, there may still be a time bomb lurking.

Occasionally the bug will be too complicated to yield to such casual troubleshooting. If the timing of a PAL will have to be adjusted, before you wildly make changes visualize the new timing in your mind or on a sheet of graph paper. Will it work? It's much faster to think out the change than to actually implement it . . . and perhaps troubleshoot it all over again.

Rapid troubleshooting is as important as accurate troubleshooting. Decide what your experiment will be, and then stop and think it through once again. What will this test really prove? I like experiments with binary results—the signal is there or it is not, or it meets specified timing or it does not—since either result gives me a direction to proceed. Binary results have another benefit: sometimes they let you skip the experiment altogether! Always think through the actions you'll take *after* the experiment is complete, since sometimes you'll find yourself taking the same path regardless of the result, making the experiment superfluous.

If the experiment is a nuisance to set up, is there a simpler approach? Hooking up 50 logic analyzer probes or digging through a million trace cycles is rather painful if you can get the same information in some easier way. I'd hate to be in a lab without a logic analyzer, since they are so useful for so many things . . . but I try to keep it as the tool of last resort, since most often it's possible to construct an easier experiment that is complete in a fraction of the time it takes to connect the LA.

Don't be so enamored of your new grand hypothesis that you miss data that might disprove it! The purpose of a hypothesis is simply to crystallize your thinking—if it is right, you'll know what step to take next. If it's wrong, collect more data to formulate yet another theory.

Step 6: *Fix the bug.*

There's more than one way to fix a problem. Hanging a capacitor on a PAL output to skew it a few nanoseconds is one way; another is to adjust the design to avoid the race condition entirely.

Sometimes a quick and dirty fix might be worthwhile to avoid getting hung up on one little point if you are after bigger game. Always, always revisit the kludge and reengineer it properly. Electronics has an unfortunate tendency to work in the engineering lab and not go wrong until the 5000th unit is built. If a fix feels bad, or if you have to furtively look over your shoulder and glue it in when no one is looking, then it *is* bad.

Finally: never, ever, fix the bug and assume it's OK because the symptom has disappeared. Apply a little common sense and scope the signals to make sure you haven't serendipitously fixed the problem by creating a lurking new one.

Speed Up by Slowing Down

There he sits . . . the organization's engineering guru, respected but somewhat feared because of his arcane knowledge. His desk is a foot deep in paper, the lab bench a mess of old food containers and smoldering solder drippings. Tools and resistor clippings threaten to short out any test system carelessly placed on the bench. Wires crisscross every square inch of tabletop—scope probes, clip leads, RS-232 cables—all going somewhere . . . though perhaps no one really knows their destination.

Ask the guru for a piece of paper and be prepared to wait. He burrows frantically through the mess. Usually the paper never comes to light. It's lost. Don't worry, though—he'll recreate it for you as soon as he has a chance. Probably the PAL equations he'll come up with will be about right, but if they're not—no problem! He's already debugged that circuit twice, so he's quite the expert.

Too many managers tolerate this level of chaos. Me, I'm a reformed lab pig. My 12-step recovery program revolved around living in tiny places—a VW microbus, many boats—which force you to be organized simply to deal with the incredible lack of living space. There's no room to be a slob on a small sailboat! Fortunately, my personal quest for organization rolled over into the lab when I discovered just how much time I saved by putting things where they belong.

Mess and clutter quite simply decrease productivity. Those few minutes a day spent putting things away save hours of searching. Sweep the solder drippings and wire segments off the bench once in a while and your incidence of catastrophic failures will plunge dramatically.

An organized lab promotes correctness. How many times have you seen engineering changes that never quite made it into production because someone forgot to write them down? Or because the notation was made on the corner of a napkin that was accidentally used to wipe up a spill and then thrown away?

When starting to debug a new project, remove everything from the bench and sweep it clean. A quick wipe with a damp cloth removes those accumulated coffee stains. Then, put everything not absolutely needed back on the shelves. This is the unique chance we get once in a while to remove the clutter, so be relentless.

Any embedded project will require at least a computer and a scope. Decide what test equipment you'll use continuously, and which will be used only on an as-needed basis. All too often even a simple embedded system has some sort of communications link requiring an extra computer as a source of data. I like to use a laptop for this as it requires little bench space.

Be sure you can easily reach the computer's frequently used connectors. If two different devices must share an RS-232 port, buy a switch box and reduce the wear and tear on connectors . . . and your back.

Don't work with unacceptable power distributions. Too many of us spend half our lives swapping power plugs. Buy outlet strips or wire up a decent source of AC mains to your test bench.

Miles and Beryl Smeaton sailed their aging boat around Cape Horn many years ago with expert boatbuilder John Guzzwell as crew. When the boat flipped in 30-foot seas and the hull cracked open, Guzzwell was shocked to discover that all of the Smeaton's tools were rusty and dull. As water poured in he carefully sharpened and cleaned the tools before undertaking the repairs that eventually saved their lives.

The moral is to buy good tools and take care of them. You'll live with those dikes and needle-nose pliers for weeks on end. Buy cheap stuff and your blood pressure will skyrocket every time you can't clip a lead close to the board. Keep them organized—get a little toolbox to keep them from falling onto the floor and getting lost.

How is your soldering equipment? A vacuum desolderer is great for making large-scale changes, but during prototyping I find it's often easier just to hack away at the board, mounting chips on top of chips and using plenty of blue wire.

During the first few days (or weeks) of bringing up a new embedded system I often find myself making lots of little modifications to the system. A hot iron always at hand is critical. After things start to more or less work, I start testing I/O interfaces by writing low-level drivers and exercising the code, making software and hardware changes in parallel as needed. The code changes much faster than the wiring, so it seems wasteful to keep an iron hot all the time. Several companies sell neat $30 cordless soldering irons that heat in seconds, the ideal thing for those infrequent modifications.

Being an immensely stupid person, I require vast quantities of clip leads. Most of my ideas are wrong, so I save a ton of time by using a clip lead to try a design change and see what happens.

Clip leads have a very short lifetime in a development lab. Accidentally connect Vcc to ground and the plastic tip melts horribly. I hate it when that happens. We used to send a runner to Radio Shack occasionally to replenish our supply but found that "the Shack" couldn't keep up with our needs.

It's better to buy 100 clips at a time and have a high-school kid solder up 50 leads. You'll have an infinite supply for a while, and may help a fledgling engineer find his true vocation. (Bring a part-timer in from your local high school to help maintain the lab. The cost is minuscule, the lab will be better off for it, and you'll show one more kid that there are alternatives to slinging burgers.)

Be sure your lab area is set up to ensure that you can also do serious software development! Clearly, your computer must include the properly installed compilers and assemblers needed for the project. Just as important as quality hand tools are the debuggers, make utilities, and other software resources needed to quickly and painlessly write, compile, and test the code. Set up the environment with a Make utility so you can compile/assemble without twiddling compiler switches.

Hardware design requires as much software support as does the firmware. PALs, PLDs, and FPGAs let you create much of the hardware design late in the game and so are a wonderful thing. Be sure your bench is set up with all of the tools you need to edit and compile these.

Documentation

All too often the frenetic pace of debugging hardware tempts us to be less than careful about writing down changes. Resist this temptation. Your company is paying you to debug a prototype for one reason only: so it can be turned into a working production system. If you carelessly forget to document modifications, the company will need at least one additional PCB revision, which you'll have to debug all over again. This is a terrible waste of money. A wise manager of such a documentation-free engineer will either retrain or fire the individual.

Avoid taking notes on scraps of paper. The best solution is a meticulously maintained engineering notebook. Write everything down, clearly and concisely. The good nuns of my grammar school all but committed suicide over their failed attempts to teach me penmanship, so such clarity is a particular headache for me. I've learned to slow down and print, since most of the time I can't read my own script.

Some engineers document directly into a computer file. If your environment is so perfect that you can always seamlessly switch to the editor, perhaps this works—if you keep backups. In most cases, though, being stuck in a program you can't exit forces you to make notes on paper.

Use *one* set of schematics to record changes. This is your master development drawing set. Staple them together and clearly label them as your masters.

When creating the schematics, go ahead and add comments, just as we do in the code. For example, document how things work.

For all off-page connections, document what page the connection goes to.

Whenever you add a part whose Vcc and GND connections are not obvious, provide a comment that indicates how power and ground connect. Power connections are as important as the logic, so someone who's troubleshooting will surely need to check these at some time. Without on-schematic notes they'll be forced to go to the databooks.

Similarly, for those nasty parts with pins protruding on all four sides, add a schematic note that indicates where pin 1 is located, and how the part is numbered (CW or CCW). Also, add tick-marks on the silk screen for every fifth pin on large parts. It makes it *so* much easier to find pin 143. . . .

Assumptions

A misspent youth of blaring rock 'n' roll left my hearing somewhat impaired, but helped formulate, of all things, my philosophy of troubleshooting digital systems. The title of the Firesign Theatre's "Everything You Know Is Wrong" album should be our modern anthem for making progress in the lab.

I hate getting called into a troubleshooting session and finding that the engineer "knows" that x, y, and z are not part of the problem at hand. *Everything you know is wrong!* Is that 5-volt supply really 5 volts at the PCB? What makes you think ground goes to the chips—when a single part has 5 or 10 ground connections, make sure *all* of them are connected. Could the system be dead because there's no clock signal? Are you *sure* the design isn't really working—could your experiment be flawed?

Another example: suppose your system runs fine at 10 MHz but never at 20. Obviously you'd put a 20-MHz clock source in and pursue the problem. Every once in a while, go back to 10 MHz just to be sure the symptom has not changed. You could spend a lot of time developing a hypothesis about 20 versus 10 operation, when the 10-MHz test results might actually be a fluke.

Assume nothing. Test everything. The PCB may have manufacturing errors on internal layers. Power and ground may not be on the pins you expect—particularly on newer high-density SMT parts. Signals labeled without an inversion bar may actually be active low. You might have ROMs mixed up. Perhaps someone loaded the wrong parts on the board.

Never blindly trust your test equipment—know how each instrument works and what its limitations are. If two signals seem impossibly skewed by 15 nsec on the logic analyzer, make sure this is not an artifact of setting it to sample too slowly. When your 100-MHz scope shows a perfectly clean logic level, remember that undetected but virulent strains of 1-nsec glitches can still be running merrily around your circuit.

When you do see a glitch, one that seems impossible given the circuit design, remember that manufacturing shorts can do strange things to signals. Is the part hot? A simple finger test may be a good short indicator.

> On its final spectacular descent to Mars in 1997, the *Mars Pathfinder* spacecraft experienced a series of watchdog time-outs. The robustly designed code recovered quickly, averting disaster.
>
> Engineers later diagnosed and fixed the code, uploading patches across 40 million miles of hostile vacuum. Interestingly enough, they found that exactly the same WDT time-outs had been noted during prelaunch testing, here on Earth. The testers had attributed the rare resets to "glitches" and ignored the problem.
>
> Now, some "glitches" have physical manifestations. In one system the timer chip went into an insane mode, where it would for no apparent reason stop outputting pulses. The problem was a reset, which I knew because only a reset—or magic (never to be discounted)—could cause the problem.
>
> The culprit was a glitch on the reset line, created by the fast logic of the emulator's pod driving the unmatched impedance of the customer's two-layer PC board. A simple resistor termination cured the problem.

On another system the processor's internal I/O lost its configuration every few minutes; all of the internal registers changed to default states, yet the program continued to run fine, though all system I/O was idled.

The culprit was again a reset glitch. In this case the pulse was created by PCB crosstalk. Only one nanosecond wide, it was too short to catch reliably on a 500-MHz logic analyzer. We sampled dozens of the erratic resets, eventually creating a statistical view of the glitch.

Though every processor has a minimum reset time at least several clocks long, even very short glitches can drive CPUs and peripherals into bizarre modes. The trick is identifying the source of the problem . . . and never ignoring erratic results or hard-to-diagnose symptoms.

Bob Pease, of analog design fame, recommends, "When things are acting funny, measure the amount of funny."

Diagnose all glitches. If the system behaves oddly, something is wrong. Find the problem, or your customer will.

Learn to Estimate

At the peril of sounding like one of the ancients, I do miss the culture of the slide rule. Though accurate answers might have been elusive, we did learn to estimate the answer for every problem before attempting a solution. Alas, it's a skill that is fading away.

Calculator abuse—computing without thinking—is now too ingrained in our society to waste effort fighting. Bummer. Other instruments, though, also tempt us to mentally coast, to *do* things without thinking. Take the scope: I can't count the times an engineer mentioned that he sees the signal . . . but has no idea, when I ask, about the width of the pulse. Is it 1 nsec? 1 μsec? Perhaps a second wide?

Timing is critical in computers, yet too many of us use the scope as a sort of logic probe. "Hey, the signal is there!" Which signal? If you expect a 10-μsec pulse every msec, then any deviation from that norm is simply wrong. *Know* what to expect, and *then* ensure that the waveforms are approximately correct. A misused scope will generate a morass of misinformation.

Estimate the performance of firmware before writing it. Sure, it's tough to know how many microseconds an as-yet-unwritten function will chew up, but you can use your general knowledge of systems to make some ballpark estimates about where problems will occur.

For example, a fast serial link might overrun a busy CPU. Estimate! A 38,400-baud link carries about 4000 characters/sec, or one character per 250 μsec. That is not a lot of time for any CPU, particularly the typical embedded 8-bitter. Your processor will be pretty busy servicing the data. If it's polled, then only heroic efforts will keep you within the 250-μsec timing margin.

Suppose you chose to implement the serial receive routine as an ISR—what is the overhead? An assembly routine to queue incoming data will need a dozen or two instructions, each of which will no doubt burn up two or three machine cycles. Surely you know roughly how long a machine cycle takes (including wait states) for your system . . . don't you? Given this information, you can get a reasonable timing estimate before writing a line of code.

Recently an engineer told me, "That initialization loop is clearly the problem." Oh yeah? He was looking for something burning up almost a second of time, when clearly, regardless of processor, 1000h memory zeroing iterations will run in a few milliseconds. Use your tools, one of which is your brain, to make sure you are addressing the *real* problems.

Recently I saw a technician troubleshooting a board that exhibited multiple problems. One chip was hot enough to fry eggs, yet he chose to work on another, "unrelated" symptom. Dumb move—surely the part was ready to self-destruct, which surely would create yet more grief for the poor tech.

Always check a bare PC board fresh from the fab for a short between Vcc and ground. Because there are so many access points for these two "nodes," they're the easiest to short. If there is a short, connect the bare board to a honking power supply and run some current through the short. You'll either blow it or you'll be able to find it using the "burn your finger" heat test. Either way, you'll locate the short.

Then, before you load all of the parts onto the PCB, think deeply about what subset of components are really needed to start testing. Load only those required. When you've got a dozen parts hanging on a bus, it's hell to find the one that asserts the wrong signal at the wrong time. It's far more efficient to load parts only as required, populating the board slowly in step with your testing, to make it easy to find the culprit in multiple-enable situations.

I like to power boards from a current-limited lab supply that has an ammeter. I look at the current from time to time to make sure I'm not doing anything expensively stupid. (And I load the power supply components first, testing that part of the circuit before adding the real logic.)

It's a good idea to be on the lookout for excessive heat, especially now that so many components are surface-mounted and tough to change when you blow them up.

All semiconductor devices generate some heat; big CPUs can produce quite a bit. A really hot device, one that you can't keep your finger on, is usually screaming for help. Excessive heat may indicate an SCR latchup condition due to ground bounce or a floating input.

Less dramatic overheating, much harder to detect without a lot of practice, often indicates a design flaw. Your finger can give important clues about the design. If two devices try to drive the bus at the same time, they'll overheat.

Be careful how you apply your personal temperature sensor. I've found that my calloused forefinger is insulated enough to protect me from bad burns when a part is unexpectedly frying. Thus, I gingerly touch each part; if it seems reasonably cool, I'll then use the much-more-sensitive back of my hand to try to determine if the chip is running hotter than it should. It's surprising how much information you can get with a little experience.

When starting out debugging a very fast system, crank the clock rate down to absurdly low levels. Fix the easy stuff—logic errors and the like—before tackling high-speed timing. Why deal with a vast ocean of troubles simultaneously?

When you do find the problem, and then make a change, sometimes the modification won't help. Before doing anything, double-check the change. Did you solder the wire to the right pin? The right IC? We tend to program ourselves to look for hard problems instead of the all-too-common simple mistakes.

Plan ahead. Think before doing. Don't try something without knowing what the possible outcomes are . . . and without having some idea what you'll do for any of those outcomes. You may find that the next step will be the same regardless of the results of the experiment. In this case, save time and do something else.

The best troubleshooters are closet chess grand masters. They think many steps ahead.

The most effective troubleshooting tool is a keen eye. With a working design, most problems stem from poor manufacturing. How many of us have spent hours troubleshooting a board, only to find a missing chip? Perhaps the wrong part is installed, or the correct one upside-down.

In smaller companies engineering is often production's backup for troubleshooting. Don't accept boards unless a technician has performed a careful visual inspection first.

Then, inspect it yourself. It's far faster to find most manufacturing defects by eye than by component-level diagnosis. Look for those missing and backwards chips. Check soldering and solder splashes.

Inspect soldering on through-hole boards using a not-terribly sharp pointer, such as an awl. Move it along every pin, using it as a guide for your eye (which will otherwise quickly tire looking at a sea of pins). Scan the board one chip at a time, working in a logical progression from one side of the board to the other. Look for unsoldered and poorly soldered pins, as well as solder splashes. If it looks bad, it is.

PC board defects are the most frustrating of all problems. Despite modern quality-control processes, they are still far too common. Keep the PCB artwork around as a reference, so you can see where the tracks run when it's time to fix a short or a design problem.

Often a new design suffers from a problem you just *know* you can cure by grounding a signal. Be wary of using a clip lead as a grounder: high-speed signals will see the lead's inductance as a high impedance. The ground end will be at ground, for sure. The signal end may not look much different than without the clip lead attached. Edges are so fast now, even in slow systems, that wires no longer act like wires. Solder a short—very short—run to ground, perhaps using a discarded resistor lead. I have found that grounding via a clip lead now only works on DC signals. Realize that a wire is not a wire, but is a complex transmission line whose characteristics will confound your common sense.

Use all of your tools. One Tektronix scope has a neat digital counter. I've used it for tough hardware/software troubleshooting problems. Unsure if an interrupt comes as often as it should? The counter will tell you without a doubt how many come along. Wondering if all interrupts get serviced? Put one counter on the interrupt line, and another on the acknowledge, and see that the values are identical.

Computer systems will crash and burn from a single event. Though digital scopes are wonderful at capturing single-shot signals, it's usually much easier to work with a problem that repeats itself, often, so you can run tests at will. A logic analyzer excels at finding these one-time problems, but most won't help much with electrical issues (say, marginal signal levels).

Always be on the lookout for ways to cause these events to repeat. For example, the easiest way to troubleshoot reset problems is to use a pulse generator to reset a dead CPU repeatedly, so you can scope the reset sequence.

Years ago we used a shortwave radio to listen to the operation of our system's code. With a little experience we knew what sort of noise to expect in each of the instrument's important operating modes. With the volume turned to a quiet murmur, any change in its buzz instantly signaled trouble. Troubleshooting is a multisensory experience. Wait! What's that? It smells like a resistor burning . . . a wire-wound, by its odor. . . . The game's afoot!

Scope Debugging

A lot of developers on a tight budget debug with a scope almost exclusively. Personally, I think this is as bad as never using one. You won't get source-level debugging, which pretty much rules it out for applications written in high-level languages.

A scope complements your tools. By itself it is inadequate; in conjunction with the rest of the toolchain it is invaluable.

Just knowing how to press the buttons is not enough. That's a little like considering yourself educated because you can recite poetry in a language you don't understand. It's important to know how and when to use the scope, and what tricks you can play to pry the maximum amount of information from buggy code.

Is your program running at all? Some embedded systems don't really do anything. They just sit quietly, monitoring some value, and produce an output only if some unlikely or infrequent event occurs. Without blinking LEDs, are you really sure the unit is alive? Sure, you can use an emulator or logic analyzer and collect trace data, but the scope provides an easier alternative. Checking for "aliveness" is the simplest scope operation, requiring the use of only a single channel and only seconds of setup time.

Though you can scope the microprocessor's data, address, and control busses, it's rather hard to decide if the CPU is running wild, or if it is doing what you'd expect. Data and address lines are notoriously ugly, even in well-behaved systems.

The best solution is to probe the chip selects to your critical I/O devices. If the code is polling these, there's a good chance it is running. If you wrote the code, you probably have a pretty good idea how often the code should go to the I/O, which gives a baseline to compare against.

The first program I write on new hardware always looks something like:

```
loop:     out     (some_port), (some_data)
          jmp     loop
```

Based on the clock rate it's easy to figure the time between OUTs. I'll scope the I/O line (whatever it is called: IORQ, M/IO, etc.), make sure the chip selects are there, and that they are spaced about right. If the system can run this loop, 90% of the time the kernel of the hardware (CPU, ROM, RAM, etc.) is functioning properly.

RS-232 is one of the biggest headaches around. It seems no serial port or routine ever works quite right at first. If you are coding a comm function that just doesn't seem to be working, use a scope to see if at least data is moving around.

Pins 2 and 3 of the RS-232 connector (for both the 9- and 25-pin versions) have the serial streams. Put a probe on each of the pins to see if there is any activity. RS-232 usually uses 12- to 15-volt levels, so be sure to crank the volts/division control to the 5- or 10-volt position. If you see no data, then the hardware or the code is broken.

Debugging serial code often involves a lot of interrupt fiddling, queue management, etc. I typically connect a scope more or less permanently to the serial lines so I'll know instantly if comm shuts down.

It pays to be a little suspicious of your hardware platform when working with early prototype systems. Being able to run a few checks yourself will saves a lot of finger pointing and aggravation, especially at 3 A.M. when your boss is screaming for results.

To a software person, the true value of a scope lies in its ability to measure the relationship between two signals. Though it's easy to apply a pair of inputs to the channel 1 and 2 vertical amplifiers, you must give some thought to setting up the scope's trigger system to get meaningful results.

Suppose your code should respond to an interrupt by driving a pattern of bits out some port, but for some reason the pattern never seems to appear. What's wrong?

Either the code never even tries to access the port, or it is sending the wrong data. Multiple causes branch from each of these possibilities, but before you can make further decisions, you'll need more information.

The first step is to look at the chip select pin on the I/O device. If it is toggling, then at least something in the software is accessing it.

Determining if the correct data is going out is a bit more difficult. If the device is one of the common ultracomplex high-integration chips, like an IEEE-488 controller, you'll have to look at the data going to it during the I/O cycle.

This is the trick to effective scope use. A data bus is always extremely busy. No one is smart enough to drop a probe on it and figure out what is going on. You must look at the bus at a particular instant in time—in this case, during the time the I/O write is in process.

In this case, put the chip select on channel 1. Use the trigger controls to trigger the scope (i.e., start the sweep) when the select comes along. Thus, select a trigger source of channel 1, and a trigger slope of (–) if the chip select goes low when it is active (usually the case). Twiddle the trigger level and time/division knobs to get a nice-looking pulse on the screen.

Now, connect the channel 2 probe to a data bus pin on the I/O device. Start with data bit 0. Look at the two signals on the CRT and note the state of channel 2 when the chip select is active. The data bus might look horrible, with ramping levels and all kinds of nonsense, but during the chip select period it will be either high or low. Note the state. Check each bit in succession, logging the pattern.

The result? You'll find out exactly what data was transferred to the device, and can use this information to shed some light on what the code must be doing.

The whole field of digital logic is based on presenting the correct data at the correct time. When you look at the confusing mess on the scope display, remember that it really doesn't matter what is up there, except during that short period of interest.

You can use this technique to add a "virtual debugging port" to any embedded system. Sometimes I'll design a system to include an extra 8-bit parallel port that drives LEDs. Then I can instrument my program to send patterns out to the displays, so I can see just what the code is doing. I'll put out a different lamp combination for each interrupt service routine, each main operating mode, etc. If things change so quickly that I can't see the LEDs blink, I watch the port with a scope.

The problem is that no boss likes to add special hardware to a system to ease debugging. One solution is to write the codes out to a nonexistent port, capturing the data on the scope instead of LEDs.

Frequently the I/O decoder has spare outputs; chip selects that were not needed. Use this unallocated "port" as the virtual debug address. Feed it into channel 1, and trigger the scope on this signal. Scope the data bus with channel 2. The I/O write to the virtual port will not affect the system, but it will give you a convenient way to trigger the scope. The data bus's contents during the write is the value your instrumented software is sending out.

Chapter 7 describes scopes in general; another very handy attribute of better oscilloscopes is delayed sweep. Just as any decent scope has at least two vertical channels, most include two time bases as well. Seems odd, doesn't it? Double vertical channels intuitively make sense, since each probe picks off a different sense point. Time, though, always flows in the same direction at the same rate, so a single axis is all that makes sense.

Novice scope users understand the operation of time base A: crank the time/division knob to the right and the signal on the screen expands in size. Rotate it to the left and the signal shrinks, but much more history (i.e., more microseconds of data) appears.

Time base B is a bit more mysterious. If enabled, it doesn't start until sometime after time base A begins. Try it on your scope: select "Both" (or "A intensified by B") and select a sweep rate faster than that used by A. You'll see a highlighted section of the trace whose width is determined by B's sweep rate, and whose starting position is a function of the delay time knob.

Switching from "Both" to "B" shows just the intensified part of the sweep: the part controlled by time base B. In effect, you've picked out and blown up a portion of the normal sweep. It's like a zoom control—and you can select the zoom factor using the sweep time, and the "pan position," or starting location, using the delay time adjustment.

Suppose you want to look at something that occurs a long time after a trigger event. Using these zoom controls you can get a very high-resolution view of that event—even when time base A is set to a very slow rate.

Delayed sweep is always accompanied by a second trigger system. Most of us have developed callouses twiddling the trigger level control in an effort to obtain stable scope displays. Any instrument with dual time bases will come with a second of these knobs to set the trigger point of the B channel.

(Note: Newer scopes, like the MSO series from HP, remove most of the uncertainty from setting trigger levels because they show an arrow on the waveform indicating the exact voltage setting of the trigger level control. It's a great time-saver.)

The second trigger is important when working on digital signals that usually have unstable time relationships. Set the A trigger to start the sweep (as always), position the intensified part of the sweep to some point *before* the section you'd like to zoom on, and then adjust trigger B until the bright portion starts exactly on the event of interest.

This procedure guarantees that even though the second trigger event moves around with relationship to trigger A, you'll see a stable scope display after selecting the B time base. In effect you've qualified trigger B by trigger A, and you can hope you're zeroing in on the area needing study.

182 THE ART OF DESIGNING EMBEDDED SYSTEMS

Delayed sweep is essential when working on any embedded system—let's look at a couple of cases.

Suppose your microprocessor crashes immediately after RESET. Traditional troubleshooting techniques call for hooking up the logic analyzer and laboriously examining all of the data and address lines. Personally, I find this to be too much trouble. Worse yet, it tends to obscure "electrical" problems: the analyzer might translate marginal ones and zeroes into what look like legal digital levels. Logic analyzers are great for purely digital problems, but any problem at power-up can easily be related to signal levels.

Only a scope gives you a view of those crucial signal levels that can cause so much trouble. Trigger channel 1 on the RESET input and probe around with channel 2. Look at READ: every processor starts off with a read cycle to grab the first instruction or startup vector. You may find a puzzling phenomenon: if the reset is provided by a source asynchronous to the processor's clock (as is the case with an RC circuit, a Vcc clamp, and even with many watchdog timers), READ will bounce around with respect to RESET. You'll never get a nice high-resolution view of READ this way.

Triggering off READ will not help. You need to catch the *first* read after reset (to look at the first instruction fetch), not any arbitrary incarnation of the signal . . . and no doubt there will be millions of reads between resets.

The answer is delayed sweep. Put RESET into the scope's external trigger input and fiddle the knobs until you get a stable trigger. (I like to put one scope channel on the external trigger while doing this initial setup to make sure the trigger is doing what I expect.) Then connect channel 1 to your processor's READ output and crank the time base until it appears over toward the right side of the display. Go to delayed (A intensified by B) mode, and rotate the B time base trigger adjustment until the bright part of the trace starts on the leading edge of the bouncing READ signal.

At this point time base A starts the sweep going on the asynchronous RESET, and time base B triggers the intensified part of the sweep when the first READ comes along. Flip the Horizontal Mode switch to B (to show only the intensified part of the sweep—that part after the B trigger), and a jitter-free READ will be on the left part of the screen. Cool, huh?

With the now stabilized scope display you can use channel 2 to look at the data lines, ROM chip selects, and other signals during the read cycle. It becomes a simple matter to see if the first instruction gets fetched correctly. A lot has to be perfect for this to happen. Very often a power-up problem comes from a bad data line, chip select, or buffer problem, any of which is trivial to find with the scope triggered properly.

This example shows how a few seconds of button twiddling can resolve two asynchronous signals on the scope display.

When your system seems crashed, it's often hard to guess exactly what the program is doing. Is the main loop running correctly? Is it stuck waiting for input from a UART?

Instead of reaching for the logic analyzer, I'll usually put on a thinking cap and speculate about what could be going on. For example, in a system that regularly polls a UART, it takes but a few seconds to check the I/O port's chip select to see if the code is hitting that pin. If so, there's a pretty good chance the main loop is at least running.

When a series of I/O operations happen sequentially you can use delayed sweep to examine each event in detail. For instance, the code to program a Zilog SCC (Serial Communications Controller—a do-everything serial link) sends many, many bytes to the same port. Triggering a scope on these port writes will display a jumble of mixed-up cycles. Delayed sweep, though, lets you trigger on the first write to the port, and then display the particular write you'd like to see.

Trigger channel A on the first write. (Use the Trigger Holdoff control to restrict triggering to burst events.) Set the sweep rate of channel B to something faster than channel A. Then use the delay time control to scroll through as many port writes as necessary to find the event causing grief. In this example, the delayed sweep lets you see a high-resolution view of events that may be widely separated in time.

Use a variation of this technique to troubleshoot many hardware/software integration issues. If your system has an unused I/O select—say, an output of an I/O decoder—seed the code with reads or writes to this port. Trigger time base A from this select, and then use delayed sweep to zoom in on an enhanced view of problem areas.

Summary—Bringing Up a New System

So there it is, your new creation, now glittering as a real bit of hardware instead of some abstract scribbles on the CAD screen. Flip on the power switch . . . and surely it'll continue staring dumbly back at you, doing nothing, dead and awaiting your magic healing touch. Whatcha gonna do?

First, before loading the parts, ohm Vcc to GND on the PCB. Any short means there's a problem with the board.

Next, load just enough parts to test the system's kernel. This includes the CPU (or maybe a socket if you're using an ICE), ROM, RAM, and decoders. Since microprocessor-based systems all use a CPU surrounded by dozens of chips all hanging on a common bus, the failure of any of which can cause problems, it makes sense to bring up your embedded system by testing the simplest sections of the hardware first.

Now stop and inspect the board carefully. Look for shorts and opens, and everything that looks a bit odd. Are all of the parts oriented properly? Are the right parts installed in the right locations? It's hell to find these sorts of problems by conventional troubleshooting techniques, so a few minutes spent inspecting may yield tremendous dividends.

Connect power, if at all possible, using a lab supply that has an ammeter. Check the meter; if it's way out of line of what you'd expect, then something serious is wrong. Stop and find the problem.

Now check the voltage and stability of Vcc on the target system. Never neglect this step, and always repeat it if weird, unexplainable things seem to be happening. A +5 supply that is even a half-volt low can cause all sorts of erratic operations that are all but impossible to troubleshoot. Check this with the scope's vertical channel on the 1 volt per division setting so you can measure the supply accurately.

Next, check the clock signal to the microprocessor. Clocks are a constant source of problems. As processor speeds increase, chip vendors are tightening specs and reducing margins. Yet even now most designers ignore the electrical characteristics of this all-important signal. If the CPU uses a crystal instead of a clock module, check the clock-out pin to make sure that it is indeed running at the correct frequency. A PCB layout problem, incorrect cut of crystal, or other problem can make the CPU start at some harmonic of the desired frequency. Again, look at this with the scope on the 1 volt per division setting so you can really see the clock's shape and voltage levels.

Test the CPU's RESET input next. This critical signal must be in an unasserted state except at power-up and reset time. If RESET is low, something is wrong.

With the basic signals correct, it's time to look at the address and data busses. You'll have two basic choices: use a tool such as an ICE or BDM, or fudge it with a bit of cleverness. Either way, check every address and data line at each chip.

Many ICEs and BDMs will let you issue a repeating write command that sends known data to all memory locations. It may be called a "Fill"; tell the tool to fill memory from 0 to infinity, over and over and over. Connect a scope to each address line and be sure that they sequence in order.

Don't have an adequate tool? Don't despair. Most CPUs include a single-byte or one-word software interrupt instruction that will serve equally well. Remove all memory chips (or disable them by putting their control signals to idle states), and pull the data bus to the value of the interrupt instruction. For example, on any x86 processor, INT3 (0xCC) is a one-byte interrupt. Z80/180 systems use RST7 (0xFF). Motorola processors usually have a breakpoint or illegal instruction trap that works equally well.

By pulling the data bus to this one-byte/word instruction, you've made it impossible for the CPU to do anything but run that particular opcode. The processor will blindly follow your will by executing the interrupt.

It will push the system context onto the stack (never doing a POP or Return), so the stack will march down to zero, and then roll over. Trigger your scope on the processor's WRITE line, and watch the addresses as the stack pointer marches along. What we've done is force the CPU to produce every possible address, in a controlled manner, while not assuming that any ROM or RAM location works!

Once the ROM works, it seems logical to assume that the code will run . . . doesn't it? Well, no. Things can and do go wrong when running code, so it makes sense to spend a few minutes trying a simple execution test before getting carried away burning complex things into ROM.

At the processor's startup location, burn the simple loop described earlier (OUT to a port, with a JMP back to the OUT) into ROM (or Flash, if you're using it). Odds are the loop will run correctly, since we've already checked the busses. Trigger a scope on the write pulse (generated by the OUT) and see that it comes at a rate correlated to your clock speed.

Next, get RAM working. Burn a bit of code that sets up the RAM chip select (if required) and that writes a location in RAM, reading the value back. With the scope, you'll be able to watch the transaction to ensure that the data comes out of RAM just as it goes in. Again, since the address bus was tested, there's no need to do an extensive test.

With working RAM and ROM, it's time to get your real software debugging tools going. If you're using a ROM monitor, build a serial port driver and link it all together. A ROM emulator should just plug in and play, now that the system's kernel is alive. An ICE or BDM, of course, will work even without an operating kernel.

Using your debugger, check the I/O using the hacking techniques outlined in Chapter 5.

CHAPTER **9**

People Musings

Managing Yourself and Others

Anyone can crank code or draw logic diagrams. Truly gifted engineers are those who predictably deliver quality products on time, on budget, that meet the specs.

Raw inspiration accounts for a tiny fraction of the effort needed to be constantly successful. An awful lot of what we do boils down to finding a reasonable formula for success and then following that formula relentlessly. Sure, we should experiment with it, tune things as needed, but disaster is guaranteed when we abandon the process and just start hammering out code and drawings.

Chapter 2 presented and described seven steps that are fundamental to getting decent products out. Sometimes it's hard to translate ideas into daily action plans. It's even more difficult to audit one's performance in the chaos of a project, one that is surely constrained to the breaking point by schedule pressures.

So here's a "Weekly Audit," a checklist the wise developer will consult to ensure that the processes are effective and actually being used. Check it weekly, perhaps every Friday morning, without fail.

As I mentioned in the very first chapter of this book, use a Daytimer or similar time management tool as an electronic nag to remind you to do the right things at the right time. Have the Daytimer pop up a reminder to run the audit weekly. Depend on memory and you'll surely forget.

Version Control System

Yes　No　Are all source code and related files managed by a networked VCS?

Yes　No　Does each developer have only those modules absolutely needed checked out (answer "no" if they hoard checked-out modules)?

Yes　No　Has the VCS been backed up every day this week? Are the backups stored in a safe place?

*If any Nos circled: What action will you take **today** to solve the problem?*

Firmware Standards

Yes　No　Is the Firmware Standards Manual the bible for all development (answer "no" if it's stored in a musty closet like a demented nephew, paraded out for show once in a while)?

Yes　No　Is *every* function and module held to the Standards Manual, as audited by Code Inspections?

Yes　No　Do all developers buy into the Standard (answer "no" if they constantly squabble over the contents of the Standard)?

Yes　No　Was every bit of code tested this week inspected first?

Yes　No　Do all Inspection teams keep and use standard forms for tracking the number and type of each defect?

Yes　No　Do the teams all use an Inspection Checklist?

Yes　No　Do all of the developers buy into the need for Code Inspections?

*If any Nos circled: What action will you take **today** to solve the problem?*

Bug Management

Yes　No　Are the developers all using engineering notebooks to control and log defects?

Yes　No　For code being tested, is every bug logged and counted?

Yes　No　Are bad modules identified and rewritten?

Yes　No　Are more than 5% of the modules falling into the "bad" category?

People Musings **189**

Yes   No   Have bug lists been abandoned (i.e., bugs fixed as they appear)?
Yes   No   For released products: is every bug being systematically tracked?

*If any Nos circled: What action will you take **today** to solve the problem?*

Tools

Yes   No   Are the development tools stable (answer "no" if they're effectively held together with baling wire and duct tape)?
Yes   No   Are all processes automated (compile, link, make, debugger initial configuration load)?
Yes   No   Does every developer have reasonable access to the tools (answer "no" if people are waiting for access)?
Yes   No   Are hand tools, clip leads, and the like in good condition?
Yes   No   Are there adequate supplies of logic analyzer clips and the like?
Yes   No   Is the "bozo" bit reset (answer "no" if anyone is doing something stupid, like holding systems together with propped-up books, or building 3-D clip-leaded prototypes that look like works of modern sculpture)?

*If any Nos circled: What action will you take **today** to solve the problem?*

Tracking Development Rates

Yes   No   Is every engineer filling out time cards accurately? (Answer "no" if this is a mad scramble at the end of the week, which indicates you'll never learn how long it takes to build a product or write a line of code.)
Yes   No   Is every diversion (such as switching to another project for a few hours) tracked?

*If any Nos circled: What action will you take **today** to solve the problem?*

Work Environment

Yes   No   Does each developer know his or her most productive time, and then use that time wisely (answer "no" if developers don't close their doors or otherwise warn off interruptions during these hours)?

Yes No Does every developer turn off the phone for at least several hours a day during their productive time?

Yes No Do developers limit time they leave their email reader on?

Yes No If cubicles are the norm, does each developer do something (e.g., wear headphones) to limit noise distractions?

*If any Nos circled: What action will you take **today** to solve the problem?*

Critical Paths

- What action can you take today to make sure everyone has what they need to be successful next week?
- What action can you take *next week* to make sure everyone has what they need to be successful next month?

Note that each category concludes with the important admonition: do something today to clear the roadblock. Don't defer action; it's much easier to correct a project when it first starts to veer off course than after months of dysfunctional development have left their scars.

Boss Management

Management is the art of combining resources in innovative ways to get a desired outcome. In our industry this outcome is some blend of features, quality, and schedule.

Yet *schedule* is the usual battleground between managers and the managed. When management distorts or destroys your careful estimate, or beats you into agreeing to one that cannot possibly happen, failure is certain. Period. Yet this practice is the norm.

People ask me constantly how they can better estimate the time a project will take. When I probe, usually I find that dates are assigned capriciously by marketing or upper management. These engineers don't really want to know how to better estimate their schedules; they're looking for a silver bullet, a bit of magic that will let them shoehorn their project into an impossible time frame. Magic and estimation are two very different things.

Bosses complain that the engineers pad their estimates so much that there simply must be fat. They feel justified in whacking off a month or two. Or, there are those who feel an aggressive schedule inspires harder work—possibly true, but only when "aggressive" is not confused with "impossible."

My feeling is that if there's no mutual trust between workers and management, the employment situation is dysfunctional and should be terminated. Professionals—us!—are paid for doing the work and for making reasonable technical recommendations. We may be wrong sometimes, but a healthy work environment recognizes the strengths and weakness of each professional. If your boss thinks you're an idiot, or refuses to trust your judgment, search the employment ads.

Too many bosses have little or no experience in managing software projects. The news they get is invariably bad—the project will take six months longer than hoped—yet it generally comes with no options, no decisions that he can make to achieve the sort of balance between product and delivery.

It's critical that we learn to manage our bosses. When presenting bad news, be sure you give options. "We can deliver on time but without these features, or 6 months late with everything, or on time but with lots of bugs. . . ." An intelligent analysis of choices, presented clearly, will help get your message across.

We need to develop trust with our superiors by educating them about development issues, by being right (meeting our own predictions), and by communicating clearly.

We've got to avoid quoting a long, arbitrary time impact as a knee-jerk reaction to any change request.

Too many developers react to a manager's request by obfuscating the facts. A schedule question gets answered with a long discourse peppered with obscure acronyms and a detailed analysis of the technology involved. In most cases your boss will not be as good as you are at cranking code or designing FPGA equations. The boss is paid to manage, not do. We're paid to do, and to communicate clearly to the rest of the organization. When talking to the boss, talk his lingo, not the language of ones and zeroes.

If we expect to be treated honestly and with respect, we have to reciprocate accordingly.

Just as it takes time and many projects to get the data you need to be an accurate estimator, educating the boss and creating trust can be a very slow process. So slow, in fact, that you must remember that sooner or later the boss will die or move on . . . and you'll be in charge. Then remember. Treat your people with trust and respect, and teach them what you've learned about scheduling.

Evolution is a great thing. Perhaps the firmware industry will mature as new generations of people learn to do things correctly, and then slowly replace the dinosaurs now all too often at the top.

Managing the Feedback Loop

The last step in most projects is the one we dread the most—assigning the blame. Who is responsible for the late delivery? Why didn't we meet the specification document? Who let costs spiral out of control?

The developers, that's who. When management sheds blame like a duck repels water, we wonder why we got into such an unforgiving profession.

Something happened in this country in the past couple of decades, something scary for the future. We've become intolerant of failure. In 1967 a horrible fire consumed the Apollo 1 spacecraft and three astronauts. An investigation found, and corrected, numerous problems. There was never a serious question about carrying on.

In the 1980s, when the *Challenger* blew up, commentators asked what NASA was doing to ensure that such a tragedy would *never* happen again. Huh? Sitting on 6 million pounds of explosive and you want a guarantee that the system was foolproof? Even my car is not totally reliable. There are no guarantees, yet society seems to expect miracles from us, the technology gurus.

Consider the Superconducting Supercollider. If scientists could promise a practical result, or perhaps only promise finally resolving the issue of the Higgs particle, then maybe the SSC would be something more than an abandoned hole in the ground. Fear of failure sent the politicians fleeing. Yes, it was very, very expensive. I was angered, though, by the national lack of understanding that, in science, failure is an element of success. We learn by trying a lot of things; with luck, a few pan out. From each defeat we have the possibility of crawling toward success.

As developers, we've got to learn to manage both failure and success. Our companies are demanding more from us every day. Downsizing and increasingly frenetic time-to-market pressures mean that Joe Engineer must take advantage of every opportunity to learn.

Yet there is no Embedded University. We're mostly educated via OJT, a haphazard and inefficient way of learning. Few of us are privileged to work with a mentor of stature, so the best we can do is to examine the results of everything we do, with a critical, unbiased eye toward improving our skills, and improving the processes used to develop our products.

Does this scenario sound familiar? A small team starts a project with great hopes and enthusiasm. Along the way problems crop up. Sales changes the features. Management reduces the product's cost. Schedules slip when compiler bugs appear. Code grows bigger than expected. Real-time response isn't adequate, so the engineers start burning the midnight oil, making heroic changes to get the system out, but schedules slip more, tempers flare, and when the product finally ships no one is speaking to each other.

A week later the developers are embroiled in another product, again starting with high hopes, and again doomed to encounter the same rather small yet common set of problems that cause late delivery.

Sliding into middle age one has the chance to observe patterns in one's life, patterns we seem to repeat over and over. Einstein said, "Doing the same things over and over, and expecting different results each time, is clearly insane."

Yet most engineering efforts exhibit this insanity. Careening from project to project, perhaps learning a little along the way but repeating the same tired old patterns, is clearly dysfunctional.

In most organizations the engineering managers are held accountable for getting the products out in the scheduled time, at a budgeted cost, with a minimal number of bugs. These are noble, important goals.

How often, though, are the managers encouraged—no, *required*—to improve the *process* of designing products?

The Total Quality movement in many companies seems to have by-passed engineering altogether. Every other department is held to the cold light of scrutiny, its processes tuned to minimize wasted effort. Engineering has a mystique of dealing with unpredictable technologies and workers immune to normal management controls. Why can't R&D be improved just like production and accounting?

Now, new technologies are a constant in this business. These technologies bring risks, risks that are tough to identify, let alone quantify. We'll always be victims of unpredictable problems.

Worse, software is very difficult to estimate. Few of us have the luxury of completely and clearly specifying a project before we start. Even fewer don't suffer from creeping featurism as the project crawls toward completion.

Unfortunately, most engineering departments use these problems as excuses for continually missing goals and deadlines. The mantra "Engineering is an art, not a science" weaves a spell that the process of development doesn't lend itself to improvement.

Phooey.

Engineering management is about removing obstacles to success. Mentoring the developers. Acquiring needed resources.

It's also about closing feedback loops. Finding and removing dysfunctional patterns of operation. Discovering new, better ways to get the work done.

Doing things the same old way is a prescription for getting the same old results.

It's infuriating that typical projects fizzle out in a last-minute crunch of bug fixes, followed by the immediate startup of a new development effort. Nothing could be dumber.

Did you learn *anything* doing the project? Did your co-workers? Is there any chance some bit of wisdom could be extracted from its successes and failures—a bit of wisdom that may save your butt in the future? Why do we careen right into the next project, hoping to avoid disaster by sheer hard work, instead of taking a moment to take a deep breath, gather our wits, and understand what we've learned?

Engineering managers simply *must* allocate time for a careful postmortem analysis of each and every project. Once the pressure of the ship date is gone, all of the team members should work toward extracting every bit of learning from the development effort.

Usually we casually pick up some wisdom even without a formal postmortem. This is the basis for "experience," a virtue acquired by making mistakes. I'll never forget shoehorning an RTOS into an almost complete system more than a decade ago. Putting it in after 20,000 lines of code were written hurt so badly I swore I'd never start a system like that again without installing an RTOS as the first software component. This bit of wisdom came in exactly the same way kids learn not to touch a hot stove: pain. I believe we can do better than learning by acquiring scars.

A formal postmortem analysis has one goal: squeeze every bit of learning from the just-completed project. Wring it dry, extracting information to compress the acquisition of "experience" as much as possible.

The postmortem is not a forum for assigning blame. When I started conducting these at my last company, the engineers immediately became paranoid, thinking that this was the chance for management to "get" them, in writing, in a venue visible to all employees.

If blame must be given, then do it privately and constructively. Nonconstructive criticism is a waste of time, to be used only when firing the offending employee (if then).

Similarly, the postmortem shouldn't be used as a staging area for the engineers' complaints against management. When there are valid concerns (for example, schedule slippages due to changing specs), then these should be coldly, accurately documented in a form that's useful to all involved. No whining allowed.

No, a successful postmortem is an unemotional, nonconfrontational, reasoned, thoughtful process. It works when all participants buy into the idea that improvement is important and possible.

I feel that a successful postmortem results in a written document that will be preserved with other engineering materials, perhaps in a drawing system. The document is important, as it's a formal analysis of ways of doing engineering better. Just as a contract is a written version of an informal understanding, the postmortem report codifies the information.

A *great* postmortem results in a report that's eminently readable, that even people not involved with the project can understand. File these together and give them to all new hires to give them "virtual experience."

The document is a critical look at every part of the project (Figure 9-1). Did the specifications change often? How often, and what was the *real* impact on the project? Were the tools up to snuff? What other toolchains could you have used, and why didn't you? Did real-time problems cause trouble? Did you badly estimate the scope of the system . . . and if so, why?

Never forget to look at the skills of all of the players. Did a new language no one really understood create problems? Perhaps new hires just didn't understand the company's technology.

Structure the report as a series of recommendations. "The tools sucked" is useless. Better: "The selected CPU had no real tool support. Next time pick a chip with at least two different ICEs and three compilers so we have options."

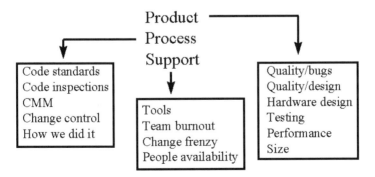

FIGURE 9-1 Areas a post mortem should cover.

A classic complaint at the end of any project is that creeping featurism inflated the spec. The post mortem must address this, in a quantitative way. No: "Marketing kept changing the specs" may be accurate, but leaves a manager no specific information useful to the next project. Better: "Four spec changes, with a total impact of 23 additional development days, accounted for 60% of the schedule slip. All changes made sense in terms of the goals. Unhappily, management forgot the impact and kept the same schedule. Next time get their approval in writing for the slip."

The goal is not to find failure, but to find answers. Successes are every bit as important to understand, so you can capitalize on them next time.

No one person is smart enough to find solutions to all problems. The document should be input to a brainstorming meeting where your colleagues hash out better ways to perform next time. Feed these ideas, where appropriate, back into the document.

The only bad post mortem is one that's not honest and thoughtful. Do assess yourselves without beating each other up—no matter how badly things went. But be intolerant of flippant, whiny, or unreflective post mortems. If a team member is unable or unwilling to look for ways to improve the organization, especially in this nonthreatening context, then that person is simply not suited to a career in this fast-changing industry. At least not with me.

A post mortem without specific quantifiable data is a waste of time. "Well, we ran somewhat late and were over budget" is useless information. "We finished early and saved a ton of money" is just as bad. You can't take action, or learn things, without knowing the specifics of the situation.

But our memories are notoriously unreliable. During a six-month project lots of things happen, good and bad. Many dates might be missed and many met. By the time you're analyzing the results of the project, there's no way you'll remember—accurately—even a few of these.

Preserve the data, so during the post mortem you'll have the accurate information you need to produce useful recommendations. The engineering notebook, which I've endorsed throughout this book, is a logical place to record all of this information.

Too many people feel that college is the end of education. It's just the start. We've all got to struggle forever to learn more and to improve. Reading, studying, seminars, trade shows are all important ingredients. Equally important is self- and organizational examination, looking for good things to emulate and bad things to fix.

Degrees

A friend went away to college at age 18, for the first time leaving home behind. A scholarship program lined his pockets with cash, enough to pay for tuition, room, and board for a full year.

A few months later he was out, expelled for nonpayment of all fees and a GPA that rivaled those of the students in *Animal House*. The money somehow turned into parties—parties that kept him a long way from class.

Today he's a successful mechanical engineer. With no degree he managed to apprentice himself to a startup, and to parlay that job into others where his skills showed through, and where enlightened bosses gave him the title and the work he's so adept at.

Over the years I've known others with similar stories, many of which ended on not-so-happy notes. The draft during the Vietnam era was, in a way, a tough burden for many smart people. They came back older, perhaps with families they had to support, and somehow never made it back to college. Many of these people became technicians, bringing their military training to a practical civilian use. Some managed to work themselves up to engineering status. Others were not so lucky.

My dad breezed through MIT on a full scholarship. Graduating with a feeling that his prestigious scholarship made him very special, he started working in aerospace. The company put him on the production line for six months, riveting airplanes together. In those days this outfit put all new engineers in production to teach them the difference between theory and practicality. He came out of it with a new appreciation for what works and for the problems associated with manufacturing. I've always thought this an especially enlightened way to introduce new graduates to the harsh realities of the physical world.

Most of today's new engineering graduates do have some experience with tools and methods. Schools now have them build things, test things, and in general act like real engineers. Still, it seems the practical aspects are subjugated to theoretical ones. You really don't know much about programming until you've completely hosed a 10,000-line project, and you know little about hardware until you've designed, built, and somehow troubleshot a complex board.

Experience is a critical part of the engineering education, one that's pretty much impossible to impart in the environment of a university. We're still much like the blacksmith of old, who started his career as an apprentice, and who ended it working with apprentices, training them over the truth of a hot fire. Book learning is very important, but in the end we're paid for what we can *do*.

In my career I've worked with lots of engineers, most with sheepskins, but many without. Both groups have had winners and losers. The non-degreed folks, though, generally come up a very different path, earning their "engineering" title only after years as a technician. This career path has a tremendous amount of value, as it's tempered in the forge of more hands-on experience than most of their BSEE-laden bosses.

Technicians are masters of making things. They are expert solderers—something far too few engineers ever master. A good tech can burn a PAL, assemble a board, and use a milling machine. The best—those bound for an engineering career—are wonderfully adept troubleshooters, masters of the scope. Since technicians spend their daily lives working intimately with circuits, some develop an uncanny understanding of electronic behavior.

Some companies won't let engineers touch a product. A tech is the developer's hands and senses. Though the engineer knows more about what the system *should* do, I imagine the techs have a deeper understanding of what it *does* do.

Too many of us view our profession parochially, somehow feeling that college is the only route to design. Part of this probably stems from the education itself, where instructors without doctorates cannot become full professors. Some comes from our fascination with honors and fancy certificates. Doctors and lawyers plaster degrees and awards over the walls to impress clients . . . which implies that we, the public, are indeed impressed by these paper honors.

These same doctors and lawyers have very effective professional associations that limit entry into the field to those people with a degree—from a school approved by the association. It's a clever way to maximize salaries through anticompetitive measures.

Electronics is very different. We're in a much younger field, where a bit of the anarchy of the Wild West still reigns. More so than in other professions, we're judged on our ability and our performance. If you can crank working designs out at warp speed, then who cares what your scholastic record shows?

And yet, our creations get more complex every day. A 1975-era embedded system pushed the edge of technology at 4 MHz, yet required little of the theoretical knowledge we got in college. One needed the ability to read a data book, the experience to know how to create circuits, and the ability to make the silly thing work.

Today's designs are different. We battle Maxwell's equations every time we propagate a fast signal more than a few inches. Our products' algorithms rely on Fourier transforms and other advanced mathematical concepts. After resisting all of the math they fed us, now I feel a little bit like the teenager coming of age—our professors, like our parents, were right after all!

Other neglected parts of a college education are becoming important. One of the most crucial: writing skills. Engineers are notoriously poor communicators, yet we're the folks building the communications age. After decades of decline, writing has assumed a new importance in the form of email. We're judged by our composition skills every time we toss off a message.

Of course, few engineering programs focus on writing. It's as if the intent is to produce development androids without the skills needed to "interface" with the rest of the world.

Occasionally we hear talk of turning engineering education into more of a vocational program. Train students to design systems and nothing else! The model fits well into the 1990s' frenetic preoccupation with getting results today, and the future be damned. If we agree that a tech, who has a VoTech-like education, could be a good engineer, then perhaps there's value to revolutionizing our schools.

Yet, I worry for the future of our profession. Several forces are shaping profound and scary changes.

The first is simply the breathtaking rate of change. Every three years or so it seems we're in a totally new sort of technology. This will only accelerate, which means the engineer of the future will either have a three-year career, or will become adept at anticipating and embracing change. More than anything, it means we have to reeducate ourselves daily.

Yet I talk to engineers every day who spend little to no time keeping current.

Time to market is another force that will change the profession. When you're designing a product, there's no time to learn how to do it, or to master the product's technology. Companies want experts *now*. Yet how can you be an expert at new technology? This is one reason we see so many consultants working in development efforts—they (effectively or otherwise) bring new knowledge to bear immediately. Enlightened management will find a way to transfer this knowledge to the core employees. Sadly, too many can't see beyond getting the product out the door, never investing in growing their skill sets.

Finally, we see a serious pigeonholing of skills. Are you good at x? Then do x! Do it forever! We can always get a new kid to work on the next project—after all, you're the x expert!

The complexity of software will only make this worse. Design a product, get it out the door, and there's a good chance you'll be involved in its maintenance forever.

You've got to take charge of your career. Manage it. Keep learning and stretching your skill set.

But I wonder how many techs-turned-engineers have the background to keep up in this rapidly advancing world. Similarly, I wonder how many college-educated designers remember enough math to understand what's going on. I did a survey recently of several graduate engineers. None could integrate a simple function. None remembered much about the transfer function of a transistor. Though these were digital folks who work with ICs, does this mean that the background and the theory drummed into them so long ago is worthless? Does it imply that only the youngest, those who haven't had time to forget, should work on the newest and the most complex systems?

I wish I knew the answer. I've tried not to discriminate on the basis of a degree, having had some wonderful experiences with very smart, very hard-working people who became engineers by the force of their will. But over time I see fewer of these. More and more résumés are filled with BS, CS, several minors, one or more masters, and the like. There's a competitive pressure that raises the stakes in job seeking. If one degree is good, we seem to think more is better.

Clearly, any large organization will screen non-degreed people out before they can demonstrate their (possibly) astonishing abilities.

Engineering is a very diverse discipline. We need thinkers and doers, inventors and implementers, designers and troubleshooters. Sometimes one person contains all of these skills, though more often a team comes together to complement each other's skills. The whole is greater than the parts.

When it's time to hire, most of us look for the standard requirements, probably including some sort of degree. I like to use the SWAN model: Smart, Works hard, Ambitious, and Nice. Though hard to gauge at an interview, these qualities almost guarantee a decent worker. When hiring a non-entry-level person, the SWAN model, coupled with what they've done in the past, is a far better indicator of success than any sheepskin.

As someone who rejects our fascination with form over substance, I think that good, non-degreed engineers are a valuable asset only a fool would reject. However, not getting a degree is clearly a mistake. One just cannot compete in the job market without this prerequisite. I know—I dropped out of college three courses short of a BSEE.

Older folks who, by circumstance or bad planning, did not complete college should look at other degree options. Check out *High Technology Degree Alternatives*, by Joel Butler (ISBN 0-912045-61-2), 1994, Professional Publications. It's full of ideas about getting a degree without quitting your job or spending a lot of money.

APPENDIX **A**

A Firmware Standards Manual

Every day we make a choice: create firmware in a consistent, repeatable way, or just crank out code as whim dictates. Though it is possible to build successful products using chaotic and ill-disciplined methods, two generations of research shows that ad hoc development ultimately results in poor code delivered late.

No firmware organization can seriously consider itself "professional" unless it has a set of standards to which *all* code is held. Those standards must be in writing and absolutely clear. Developers must buy into the concept of using standards—or be retrained or replaced. Period. Code inspections insure every bit of software is audited to the standard.

Use the following standard intact, or modify it to suit your requirements. Feel free to download the machine-readable version from www.ganssle.com/ades/fsm.html.

Scope

This document defines the standard way all programmers will create embedded firmware. Every programmer is expected to be intimately familiar with the Standard, and to understand and accept these requirements. All consultants and contractors will also adhere to this Standard.

The reason for the Standard is to insure all company-developed firmware meets minimum levels of readability and maintainability. Source code has two equally important functions: it must *work*, and it must clearly *communicate how it works* to a future programmer or the

203

future version of yourself. Just as standard English grammar and spelling make prose readable, standardized coding conventions ease readability of one's firmware.

Part of every code review is to insure the reviewed modules and functions meet the requirements of the Standard. Code that does not meet this Standard will be rejected.

We recognize that no Standard can cover every eventuality. There may be times where it makes sense to take exception to one or more of the requirements incorporated in this document. Every exception must meet the following requirements:

- *Clear Reasons*—Before making an exception to the Standard, the programmer(s) will clearly spell out and understand the reasons involved, and will communicate these reasons to the project manager. The reasons must involve clear benefit to the project and/or company; stylistic motivations, or programmer preferences and idiosyncrasies are not adequate reasons for making an exception.
- *Approval*—The project manager will approve all exceptions made.
- *Documentation*—The effected module or function will have the exception clearly documented in the comments, so during code reviews and later maintenance, the current and future technical staff understand the reasons for the exception, and the nature of the exception.

Projects

Directory Structure

To simplify use of a version control system, and to deal with unexpected programmer departures and sicknesses, every programmer involved with each project will maintain identical directory structures for the source code associated with the project.

The general "root" directory for a project takes the form:
/projects/project-name/rom_name
where

- "/projects" is the root of all firmware developed by the company. By keeping all projects under one general directory, version control and backup are simplified and reduce the size of the computer's root directory.
- "/project-name" is the formal name of the project under development.

- "/rom_name" is the name of the ROM the code pertains to. One project may involve several microprocessors, each of which has its own set of ROMs and code. Or a single project may have multiple binary images, each of which goes into its own set of ROMs.

Required directories:

/projects/project-name/tools—compilers, linkers, assemblers used by this project. All tools will be checked into the VCS so in 5 to 10 years, when a change is required, the (now obsolete and unobtainable) tools will still be around. It's impossible to recompile and retest the project code every time a new version of the compiler or assembler comes out; the only alternative is to preserve old versions, forever, in the VCS.

/projects/project-name/rom_name/headers—all header files, such as .h or assemble include files, go here.

/projects/project-name/rom_name/source—source code. This may be further broken down into header, C, and assembly directories. The MAKE files are also stored here.

/projects/project-name/rom_name/object—object code, including compiler/assembler objects and the linked and located binaries.

/projects/project-name/rom_name/test—This directory is the one, and only one, that is not checked into the VCS and whose subdirectory layout is entirely up to the individual programmer. It contains work-in-progress, which is generally restricted to a single module. When the module is released to the VCS or the rest of the development team, the developer must clean out the directory and eliminate any file that is duplicated in the VCS.

Version File

Each project will have a special module that provides firmware version name, version date, and part number (typically the part number on the ROM chips). This module will list, in order (with the newest changes at the top of the file), all changes made from version to version of the released code.

Remember that the production or repair departments may have to support these products for years or decades. Documentation gets lost and ROM labels may come adrift. To make it possible to correlate problems to ROM versions, even after the version label is long gone, the version file should generate only one bit of "code"—a string that indicates, in ASCII, the current ROM version. Some day in the future a technician—or yourself!—may then be able to identify the ROM by dumping the ROM's contents. An example definition is:

```
# undef VERSION
# define VERSION "Version 1.30"
```

Example:

```
/******************************************************
* Version Module—Project SAMPLE
*
* Copyright 1997 Company
* All Rights Reserved
*
* The information contained herein is confidential
* property of Company. The use, copying, transfer
* or
* disclosure of such information is prohibited
* except
* by express written agreement with Company.
*
# undef VERSION
# define VERSION "Version 1.30"
* 12/18/97—Version 1.3—ROM ID 78-130
*               Modified module AD_TO_D to fix
*               scaling
*               algorithm; instead of y=mx, it
*               now
*               computes y=mx+b.
* 10/29/97—Version 1.2—ROM ID 78-120
*               Changed modules DISPLAY_LED and
*               READ_DIP
*               to incorporate marketing's
*               request for a
*               diagnostics mode.
* 09/03/97—Version 1.1—ROM ID 78-110
*               Changed module ISR to properly
*               handle
*               non-reentrant math problem.
* 07/12/97—Version 1.0—ROM ID 78-100
*               Initial release
******************************************************/
```

Make and Project Files

Every executable will be generated via a MAKE file, or the equivalent supported by the tool chain selected. The MAKE file includes all of the information needed to automatically build the entire ROM image. This includes compiling and assembling source files, linking, locating (if needed), and whatever else must be done to produce a final ROM image.

An alternative version of the MAKE file may be provided to generate debug versions of the code. Debug versions may include special diagnostic code, or might have a somewhat different format of the binary image for use with debugging tools.

In integrated development environments (like Visual C++) specify a PROJECT file that is saved with the source code to configure all MAKE-like dependencies.

In no case is any tool *ever* to be invoked by typing in a command, as invariably command line arguments "accumulate" over the course of a project . . . only to be quickly forgotten once version 1.0 ships.

Startup Code

Most ROM code, especially when a C compiler is used, requires an initial startup module that sets up the compiler's runtime package and initializes certain hardware on the processor itself, including chip selects, wait states, etc.

Startup code generally comes from the compiler or locator vendor, and is then modified by the project team to meet specific needs of the project. It is invariably compiler- and locator-specific. Therefore, the first modification made to the startup code is an initial comment that describes the version numbers of all tools (compiler, assembler, linker, and locator) used.

Vendor-supplied startup code is notoriously poorly documented. To avoid creating difficult-to-track problems, *never* delete a line of code from the startup module. Simply comment out unneeded lines, being careful to put a note in that you were responsible for disabling the specific lines. This will ease re-enabling the code in the future (for example, if you disable the floating point package initialization, one day it may need to be brought back in).

Many of the peripherals may be initialized in the startup module. Be careful when using automatic code generation tools provided by the processor vendor (tools that automate chip select setup, for example). Since many processors boot with RAM chip selects disabled, always include the chip select and wait state code in-line (not as a subroutine). Be careful to initialize these selects at the very top of the module, to allow future subroutine calls to operate, and since some debugging tools will not operate reliably until these are set up.

Stack and Heap Issues

Always initialize the stack on an *even* address. Resist the temptation to set it to a odd value like 0xffff, since on a word machine an odd stack will cripple system performance.

Since few programmers have a reasonable way to determine maximum stack requirements, always assume your estimates will be incorrect. For each stack in the system, make sure the initialization code fills the entire amount of memory allocated to the stack with the value 0x55. Later, when debugging, you can view the stack and detect stack overflows by seeing no blocks of 0x55 in that region. Be sure, though, that the code that fills the stack with 0x55 automatically detects the stack's size, so a late night stack size change will not destroy this useful tool.

Embedded systems are often intolerant of heap problems. Dynamically allocating and freeing memory may, over time, fragment the heap to the point that the program crashes due to an inability to allocate more RAM. (Desktop programs are much less susceptible to this as they typically run for much shorter periods of time.)

So, be wary of the use of the malloc() function. When using a new tool chain examine the malloc function, if possible, to see if it implements garbage collection to release fragmented blocks (note that this may bring in another problem, as during garbage collection the system may not be responsive to interrupts). *Never* blindly assume that allocating and freeing memory is cost- or problem-free.

If you chose to use malloc(), *always* check the return value and safely crash (with diagnostic information) if it fails.

When using C, if possible (depending on resource issues and processor limitations), always include Walter Bright's MEM package (www.snippets.org/mem.txt) with the code, at least for the debugging.

MEM provides:

- ISO/ANSI verification of allocation/reallocation functions
- Logging of all allocations and frees
- Verifications of frees
- Detection of pointer over- and under-runs

- Memory leak detection
- Pointer checking
- Out of memory handling

Modules

General

A *Module* is a single file of source code that contains one or more functions or routines, as well as the variables needed to support the functions.

Each module contains a number of *related* functions. For instance, an A/D converter module may include all A/D drivers in a single file. Grouping functions in this manner makes it easier to find relevant sections of code, and allows more effective encapsulation.

Encapsulation—hiding the details of a function's operation, and keeping the variables used by the function local—is absolutely essential. Though C and assembly language don't explicitly support encapsulation, with careful coding you can get all of the benefits of this powerful idea as do people using OOP languages.

In C and assembly language you can define all variables and RAM inside the modules that use those values. Encapsulate the data by defining each variable for the scope of the functions that use these variables only. Keep them private within the function, or within the module, that uses them.

Modules tend to grow large enough that they are unmanageable. Keep module sizes under 1000 lines to insure tools (source debuggers, compilers, etc.) are not stressed to the point they become slow or unreliable, and to ease searching.

Templates

To encourage a uniform module look and feel, create module templates named "module_template.c" and "module_template.asm," stored in the source directory, that become part of the code base maintained by the VCS. Use one of these files as the base for all new modules. The module template includes a standardized form for the header (the comment block preceding all code), a standard spot for file includes and module-wide declarations, function prototypes and macros. The templates also include the standard format for functions.

Here's the template for C code:

```
/*****************************************************
* Module name:
*
* Copyright 1997 Company as an unpublished work.
* All Rights Reserved.
*
* The information contained herein is confidential
* property of Company. The use, copying, transfer
* or
* disclosure of such information is prohibited
* except
* by express written agreement with Company.
*
* First written on xxxxx by xxxx.
*
* Module Description:
* (fill in a detailed description of the module's
* function here).
*
*****************************************************/
/* Include section
* Add all #includes here
*
*****************************************************/
/* Defines section
* Add all #defines here
*
*****************************************************/
/* Function Prototype Section
* Add prototypes for all functions called by this
* module, with the exception of runtime routines.
*
*****************************************************/
```

The template includes a section defining the general layout of functions, as follows:

```
/*****************************************************
* Function name    : TYPE foo(TYPE arg1, TYPE arg2)
* returns          : return value description
```

```
*   arg1              : description
*   arg2              : description
*   Created by        : author's name
*   Date created      : date
*   Description       : detailed description
*   Notes             : restrictions, odd modes
***********************************************/
```

The template for assembly modules is:

```
;***********************************************
; Module name:
;
; Copyright 1997 Company as an unpublished work.
; All Rights Reserved.
;
; The information contained herein is confidential
; property of Company. The use, copying, transfer
; disclosure of such information is prohibited
; except by express written agreement with Company.
;
; First written on xxxxx by xxxx.
;
; Module Description:
; (fill in a detailed description of the module
; here).
;
;***********************************************
; Include section
; Add all "includes" here
;***********************************************
```

The template includes a section defining the general layout of functions, as follows:

```
;***********************************************
; Routine name    : foobar
; returns         : return value(s) description
; arg1            : description of arguments
; arg2            : description
```

```
; Created by          : author's name
; Date created        : date
; Description         : detailed description
; Notes               : restrictions, odd modes
;*****************************************************
```

Module Names

Though long module names are a wonderful aid to identifying what-goes-where, all too many compilers and debuggers don't properly handle names longer than 8 characters. In some cases this may be a fault inherent in the object file format or a debugging file. Limit names to 8 characters or less.

Never include the project's name or acronym as part of each module name. It's much better to use separate directories for each project.

Big projects may require many dozens of modules; scrolling through a directory listing looking for the one containing function main() can be frustrating and confusing. Therefore store function main() in a module named main.c or main.asm.

File extensions will be:

C Source Code	FileName.c
C Header File	FileName.h
Assembler files	FileName.asm
Assembler include files	FileName.inc
Object Code	FileName.obj
Libraries	FileName.lib
Shell Scripts	FileName.bat
Directory Contents	README
Build rules for make	Project.mak

Variables

Names

Regardless of language, use long names to clearly specify the variable's meaning. If your tools do not support long names, get new tools.

Separate words within the variables by underscores. Do not use capital letters as separators. Consider how much harder IcantReadThis is on the eyes versus I_can_read_this.

The ANSI C specification restricts the use of names that begin with an underscore and either an uppercase letter or another underscore (_[A-Z_][0-9A-Za-z_]). Much compiler runtime code also starts with leading underscores. To avoid confusion, never name a variable or function with a leading underscore.

These names are also reserved by ANSI for its future expansion:

E[0-9A-Z][0-9A-Za-z]*	errno values
is[a-z][0-9A-Za-z]*	Character classification
to[a-z][0-9A-Za-z]*	Character manipulation
LC_[0-9A-Za-z_]*	Locale
SIG[_A-Z][0-9A-Za-z_]*	Signals
str[a-z][0-9A-Za-z_]*	String manipulation
mem[a-z][0-9A-Za-z_]*	Memory manipulation
wcs[a-z][0-9A-Za-z_]*	Wide character manipulation

Global Variables

All too often C and especially assembly programs have one huge module with all of the variable definitions. Though it may seem nice to organize variables in a common spot, the peril is these are all then global in scope. Global variables are responsible for much undebuggable code, reentrancy problems, global warming, and male pattern baldness. Avoid them!

Real time code may occasionally require a few—and only a few— global variables to insure reasonable response to external events. Every global variable must be approved by the project manager.

When globals are used, put all of them into a single module. They are so problematic that it's best to clearly identify the sin via the name globals.c or globals.asm.

Portability

Don't assume that the address of an int object is also the address of its least-significant byte. This is not true on big-endian machines. Thus, don't make the following mistake:

```
int c;
while ((c = getchar()) != EOF)
write(file_descriptor, &c, 1);
```

Functions

Regardless of language, *keep functions small!* The ideal size is less than a page; in no case should a function ever exceed two pages. Break large functions into several smaller ones.

The only exception to this rule is the very rare case where real time constraints (or sometimes stack limitations) mandate long sequences of in-line code. The project manager must approve all such code . . . but first look hard for a more structured alternative!

Explicitly declare every parameter passed to each function. Clearly document the meaning of the parameter in the comments.

Define a prototype for every called function, with the exception of those in the compiler's runtime library. Prototypes let the compiler catch the all-too-common errors of incorrect argument types and improper numbers of arguments. They are cheap insurance.

In general, function names should follow the variable naming protocol. Remember that functions are the "verbs" in programs—they *do* things. Incorporate the concept of "action words" into the variables' names. For example, use "read_A/D" instead of "A/D_data," or "send_to_LCD" instead of "LCD_out."

Interrupt Service Routines

ISRs, though usually a small percentage of the code, are often the hardest bits of firmware to design and debug. Crummy ISRs will destroy the project schedule!

Decent interrupt routines, though, require properly designed hardware. Sometimes it's tempting to save a few gates by letting the external device just toggle the interrupt line for a few microseconds. This is unacceptable. Every interrupt must be latched until acknowledged, either by the processor's interrupt-acknowledge cycle (be sure the hardware acks the proper interrupt source), or via a handshake between the code and the hardware.

Use the non-maskable interrupt only for catastrophic events, like the apocalypse or imminent power failure. Many tools cannot properly debug NMI code. Worse, NMI is guaranteed to break non-reentrant code.

If at all possible, design a few spare I/O bits in the system. These are tremendously useful for measuring ISR performance.

Keep ISRs short! Long (too many lines of code) and slow are the twins of ISR disaster. Remember that *long* and *slow* may be disjoint; a five-line ISR with a loop can be as much of a problem as a loop-free 500-line routine. When an ISR grows too large or too slow, spawn another task and exit. Large ISRs are a sure sign of a need to include an RTOS.

Budget time for each ISR. Before writing the routine, understand just how much time is available to service the interrupt. Base all of your coding on this, and then *measure* the resulting ISR performance to see if you met the system's need. Since every interrupt competes for CPU resources, that slow ISR that works is just as buggy as one with totally corrupt code.

Never allocate or free memory in an ISR unless you have a clear understanding of the behavior of the memory allocation routines. Garbage collection or the ill-behaved behavior of many runtime packages may make the ISR time non-deterministic.

On processors with interrupt vector tables, fill every entry of the table. Point those entries not used by the system to an error handler, so you've got a prayer of finding problems due to incorrectly programmed vectors in peripherals.

Though non-reentrant code is always dangerous in a real-time system, it's often unavoidable in ISRs. Hardware interfaces, for example, are often non-reentrant. Put all such code as close to the beginning of the ISR as possible, so you can then re-enable interrupts. Remember that as long as interrupts are off the system is not responding to external requests.

Comments

Code *implements* an algorithm; the comments *communicate* the code's operation to yourself and others. Adequate comments allow you to understand the system's operation without having to read the code itself.

Write comments in *clear English*. Use the sentence structure Miss Grandel tried to pound into your head in grade school. Avoid writing the Great American Novel; be concise yet explicit . . . but be complete.

Avoid long paragraphs. Use simple sentences: noun, verb, object. Use active voice: "Start_motor actuates the induction relay after a 4 second pause." Be complete. Good comments capture everything important about the problem at hand.

Use proper case. Using all caps or all lowercase simply makes the comments harder to read.

Enter comments in C at block resolution and when necessary to clarify a line. Don't feel compelled to comment each line. It is much more natural to comment groups of lines which work together to perform a macro function. However, never assume that long variable names create "self-documenting code." Self-documenting code is an oxymoron, so add comments where needed to make the firmware's operation crystal clear. It should be possible to get a sense of the system's operation by reading only the comments.

Explain the meaning and function of every variable declaration. Every single one. Explain the return value, if any. Long variable names are merely an *aid* to understanding; accompany the descriptive name with a deep, meaningful, prose description.

Comment assembly language blocks and any line that is not crystal clear. The worst comments are those that say "move AX to BX" on a MOV instruction! Reasonable commenting practices will yield about one comment on every other line of assembly code.

Though it's useful to highlight sections of comments with strings of asterisks, never have characters on the right side of a block of comments. It's too much trouble to maintain proper spacing as the comments later change. In other words, this is not allowed:

```
/*********************************************
 * This comment incorrectly uses right-hand *
 * asterisks                                *
 *********************************************/
```

The correct form is:

```
/*********************************************
 * This comment does not use right-hand
 * asterisks
 *********************************************/
```

Coding Conventions

General

No line may ever be more than 80 characters.

Don't use absolute path names when including header files. Use the form `#include <module/name>` to get public header files from a standard place.

Never, ever use "magic numbers." Instead, first understand where the number comes from, then define it in a constant, and then document your understanding of the number in the constant's declaration.

Spacing and Indentation

Put a space after every keyword, unless a semicolon is the next character, but never between function names and the argument list.

Put a space after each comma in argument lists and after the semicolons separating expressions in a for statement.

Put a space before and after every binary operator (like +, –, etc.). Never put a space between a unary operator and its operand (e.g., unary minus).

Put a space before and after pointer variants (star, ampersand) in declarations. Precede pointer variants with a space, but have no following space, in expressions.

Indent C code in increments of two spaces. That is, every indent level is two, four, six, etc. spaces. Indent with spaces, *never tabs*.

Always place the # in a preprocessor directive in column 1.

C Formatting

Never nest IF statements more than two deep; deep nesting quickly becomes incomprehensible. It's better to call a function, or even better to replace complex IFs with a SWITCH statement.

Place braces so the opening brace is the last thing on the line, and place the closing brace first, like:

```
if (result > a_to_d) {
  do a bunch of stuff
}
```

Note that the closing brace is on a line of its own, except when it is followed by a continuation of the same statement, such as:

```
do {
  body of the loop
} while (condition);
```

When an `if-else` statement is nested in another `if` statement, always put braces around the `if-else` to make the scope of the first `if` clear.

When splitting a line of code, indent the second line like this:

```
function(float arg1, int arg2, long arg3,
         int arg4)
```

or,

```
if (long_variable_name && constant_of_some_sort ==
2
    && another_condition)
```

218 THE ART OF DESIGNING EMBEDDED SYSTEMS

Use too many parentheses. Never let the compiler resolve precedence; explicitly declare precedence via parentheses.

Never make assignments inside `if` statements. For example, don't write:

```
if ((foo = (char *) malloc(sizeof *foo)) == 0)
   fatal("virtual memory exhausted");
```

instead, write:

```
foo = (char *) malloc(sizeof *foo);
if (foo == 0)
   fatal("virtual memory exhausted")
```

If you use `#ifdef` to select among a set of configuration options, add a final `#else` clause containing an `#error` directive so that the compiler will generate an error message if none of the options has been defined:

```
#ifdef sun
#define USE_MOTIF
#elif hpux
#define USE_OPENLOOK
#else
#error unknown machine type
#endif
```

Assembly Formatting

Tab stops in assembly language are as follows:

- Tab 1: column 8
- Tab 2: column 16
- Tab 3: column 32

Note that these are all in increments of 8, for editors that don't support explicit tab settings. A large gap—16 columns—is between the operands and the comments.

Place labels on lines by themselves, like this:

```
label:
        mov    r1, r2                ; r1=pointer to I/O
```

Precede and follow comment blocks with semicolon lines:

```
;
; Comment block that shows how comments stand
; out from the code when preceded and followed by
; "blank" lines.
;
```

Never run a comment between lines of code. For example, do not write like this:

```
mov    r1, r2        ; Now we set r1 to the value
add    r3, [data]    ; we read back in read_ad
```

Instead, use either a comment block, or a line without an instruction, like this:

```
mov    r1, r2        ; Now we set r1 to the value
                     ; we read back in read_ad
add    r3, [data]
```

Be wary of macros. Though useful, macros can quickly obfuscate meaning. Do pick very meaningful names for macros.

Tools

Computers

Do all PC-hosted development on machines running Windows 95 or NT only, to insure support for long file names, and to give a common OS between all team members.

If development under a DOS environment is required, do it in a Win 95/NT DOS window.

Maintain every bit of code under a version control system. In addition, the current compiler, assembler, linker, locator (if any) and debugger(s) will be checked into the VCS. Products have lifetimes measured in years or even decades, while tools tend to last months at best before new versions appear. It's impossible to recompile and retest all of the product code just because a new compiler version is out, so you've got to save the toolchain, under VCS lock and key.

The only downside of including tools in the VCS files is the additional disk space required. Disks are cheap; when more free space is required simply buy a larger disk. It's false economy to limp by with inadequate disk space.

Compilers et al.

Leave all compiler, assembler, and linker warnings and error messages enabled. The module is unacceptable until it compiles cleanly, with no errors or warning messages. In the future a warning may puzzle a programmer, wasting time as he attempts to decide if it's important.

Write all C code to the ANSI standard. Never use vendor-defined extensions, which create problems when changing compilers.

Never, ever, change the language's syntax or specification via macro substitutions.

Debugging

You have a choice: plan for bugs before writing the code, and build a debuggable product, or (surprise!) find bugs during test in a system that is impossible or difficult to troubleshoot. Expect bugs, and be bug-proactive in your design.

If at all possible, in all systems with a parts cost over a handful of dollars, allocate at least two, preferably more, parallel I/O bits to troubleshooting. Use these bits to measure ISR time (set one high on ISR entry and low on exit; measure time high on a scope), time consumed by other functions, idle time, and even entry/exit to functions.

If possible, include a spare serial port in the design. Then add a monitor—preferably a commercial product, but at least a low-level monitor that gives you some access to your code and hardware.

Debugging tools are notoriously problematic—unreliable, buggy, with long repair times. As CPU speeds increase the problems increase. Yet these tools are indispensable. Select a dual, complementary, debugging toolchain: perhaps an emulator and a monitor. Or an emulator and a background debugger. Be sure that both sets of tools use a common GUI. This will minimize the time needed to switch between tools, and will insure there will be no file conversion problems (debuggers use many hundreds of incompatible debug file formats).

When selecting tools, evaluate the following items:

- Support—is the vendor responsive and knowledgeable? Is the vendor likely to be around in a few months or years? If the unit fails, what is the guaranteed repair time?
- Intrusion—how much does the tool intrude on the system's operation? What is the impact on debugging strategies and development time?

- Does the tool run at full target speed, or will you have to slow things down? What is the impact?
- Will the mechanical connection between the tool and the target be reliable? It's quite tough to get a decent connection to many modern SMT and BGA processors.
- Interrupts/DMA—Will the tool let you debug ISRs? Are interrupts/DMA ever disabled unexpectedly? If the tool does not respond to interrupts/DMA when stopped at a breakpoint (very common), will this have a deleterious effect on your debugging?
- Tasking—If the product uses an RTOS, the tool must provide some support for that RTOS. Insure that the debugger itself is aware of the RTOS, and can display important task constructs in a high-level format. What happens if you set a breakpoint on a task do the others continue to run? If not, what impact will this have on your development?
- Internal peripherals—Is the tool aware of the CPU's internal peripherals? Many are; they let you look at the function of the peripherals at a very high level. Do timers stop running at a breakpoint (common)? Will this cause development problems?

Be wary of doing all of your development with the tool's downloader. Burn a ROM from time to time to make sure the code itself runs properly from ROM, and to insure the product properly addresses the ROMs.

Leave all debugging resources in the product when it ships. Disable them via a software flag so they lie latent, ready for action in case of a problem. Remember the *Mars Pathfinder:* JPL diagnosed and fixed a priority inversion bug while the unit was on Mars, using the RTOS's trace debug feature, which had been left in the product.

APPENDIX B

A Simple Drawing System

Just as firmware standards give a consistent framework for creating and managing code, a drawing system organizes hardware documentation. Most middle- to large-sized firms have some sort of drawing system in place; smaller companies, though, need the same sort of management tool.

Use the following standard intact or modified to suit your requirements. Feel free to download the machine-readable version from www.ganssle.com/ades/dwg.html.

Scope

This document describes a system that:

- guarantees everyone has, and uses, accurate engineering documents.
- manages storage of such documents and computer files to make their backup easy and regular.
- manages the current configuration of each product.

The system outlined is primarily a method to describe exactly what goes into each product through a system of drawings. A top-level configuration drawing points to lower-level drawings, each of which points to specific parts and/or even lower-level drawings. After following the "pointer chain" all the way down to the lowest level, one will have access to:

- Complete assembly drawings including mod lists.
- A complete parts list.

- By reference, to other engineering documents like schematics and source files.

The system works through a network of Bills of Materials (BOMs), each of which includes the pointers to other drawings, or the part numbers of bit pieces to buy and build.

Our primary goal is to build and sell products, so the drawing system is tailored to give production all of the information needed to manufacture the latest version of a product. However, keeping in mind that we must maintain an auditable trail of engineering support information, the system always contains a way to access the latest such information.

Drawings and Drawing Storage
Definitions

The term "drawing" includes any sort of documentation required to assemble and maintain the products. Drawings can include schematics, BOMs, assembly drawings, PAL and code source files, etc.

A "Part" is anything used to build a product. Parts include bit pieces like PC boards and chips, and may even include programmed PALs and ROMs. A part may be described on a drawing by a part number (like 74HCT74), or by a drawing number (in the case of something we build or contract to build).

Drawing Notes

Every drawing has a drawing number associated with it. This number is organized by product series, as follows:

Company documentation:	#0001 to #0499
Configuration drawings:	#0500 to #0999
Product line "A":	#1000 to #1999
Product line "B":	#2000 to #2999
Product line "C":	#3000 to #3999

Every drawing has a revision letter associated with it, and marked clearly upon it. Revision letters start with the letter 'A' and proceed to 'Z'. If there are more than 26 revisions, after 'Z' comes 'AA', then 'AB', etc.

The first release of any drawing is to be marked revision 'A'. There are to be no drawings with no revision letters.

Every drawing will have the date of the revision clearly marked upon it, with the engineer's initials or name.

Every drawing will have a master printed out and stored in the MASTERs file. The engineer releasing the drawing or the revision will stamp the Master with a red MASTER stamp, and will fill in a date field on that stamp.

Though in many cases both electronic and paper copies of drawings (like for a schematic) exist, the paper copy is always considered the MASTER.

Drawing numbers are always four-digit numerics, prefixed by the "#" character.

Storage

All Master drawings and related documentation will be stored in the central repository. Master computer files will be stored on network drive in a directory (described later).

Everyone will have access to Master drawings and files. These are to be used for reference only; no one may take a Master drawing from the central repository for any purpose except for the following:

Drawings may be removed to be photocopied. They must be returned immediately (within 30 minutes) to the central repository.

Drawings may be removed by an engineer for the sole reason of updating them, to incorporate ECOs or otherwise improve their accuracy. However, drawings may be removed only if they will be immediately updated; you may not pull a Master and "forget" about it for a few days. It is anticipated that, since most of our drawings are generated electronically, a master will usually just be removed and replaced by a new version. See "Obsolete Drawings" for rules regarding the disposition of obsoleted drawings.

Artwork may be removed to be sent out for manufacturing. However, all POs sent to PC vendors must require "return of artwork and all films." He who pulls the artwork or film is responsible to see that the PO has this information. Returned art must be immediately refiled.

All drawings will be stored in file folders in a "Master Drawing" file cabinet. Those that are too big to store (like D size drawings) will be folded. Drawings will be filed numerically by drawing number.

Artwork will be stored in a flatfile, stored within their protective paper envelopes. Every piece of artwork and film will have a drawing number and revision marked on both the art/film, and on the envelope. If it is not convenient to make the art marking electronically, then use a magic marker.

Storage—Obsoleted Drawings

Every Master drawing that is obsoleted will be removed from the current Master file and moved to an Obsolete file. Obsoleted drawings will be filed numerically by drawing number. Where a drawing has been obsoleted more than once, each old version will be substored by version letter.

The Master will be stamped with a red OBSOLETE stamp. Enter the date the drawing is canceled next to the stamp. Thus, every Obsolete drawing will have two red stamps: MASTER (with the original release date) and OBSOLETE (with the cancellation date).

If old ECOs are associated with the Obsoleted drawing, be sure they remain attached to it when it is moved to the Obsolete file.

Obsoleted artwork and films will be immediately destroyed.

Sometimes one makes a small modification to a Master drawing to incorporate an ECO—say, if a hand-drawn PC board assembly drawing changes slightly. In this case duplicate the Master before making the change, stamp the duplicate OBSOLETE, and file the duplicate.

The reason for saving old drawings is to preserve historical information that might be needed to update/fix an old unit.

Master Drawing Book

Whenever a drawing is released or updated, the Master Drawing Book will be modified by the releasing engineer to reflect the new information.

The Master Drawing Book is a looseleaf binder stored and kept with the Master drawing file. The Master Drawing Book lists every drawing we have by number and its current revision level. In addition, if one or more ECOs is current against a drawing, it will be listed along with a brief one-line description of what the ECO is for.

Just as important, the Master Drawing Book lists the name of the electronic version of a drawing. This name is always the name of the file(s) on the network drive, with the associated directory path listed.

Note that the "Dash Number" (described later under "Bills of Materials") is not included in the list, since one drawing might have many dash numbers.

Thus, the drawing list looks like:

Dwg #	Revision	Rev date	Title	Filename
#1000	A	8-1-97	Prod A BOM	PRODA-ASSY
ECO: PRODA.A.3		Stabilize clock		PRODA\ECO.A
ECO: PRODA.A.1		Secure cables		PRODA\ECO.A
#1001	B	8-2-97	Prod A Baseplate	PRODA-BASE

As drawings are updated the ECOs will no longer apply, and should then be removed from the book.

Note that after each BOM drawing number there is a list of dash numbers that describe what each configuration of the drawing is.

A section at the end of the book will contain descriptions of "Specials"—units we do something weird to to make a customer happy. If we give someone a special PAL, document it with the source code and notes about the unit's serial number, date, etc. A copy of this goes in the unit's folder. It is the responsibility of the technician to insure that the folder and Master Drawing Book are updated with "special" information.

The Master Drawing Book master copy will be stored as file name ENGINEER\DOCS\MDB.DOC, and is maintained in Word.

Configuration Drawings

Every product will have a Configuration Drawing associated with it. These Drawings essentially identify what goes into the shipping box.

Currently, the following Configuration Drawings should be supported:

Dwg #	Description
#0501	Product A
-1	256k RAM option
-2	1 Mb RAM option
-3	50 MHz option
#0502	Product B
#0503	Product C
-1	256k RAM option
-2	1 Mb RAM option
-3	50 MHz option

The "dash numbers" are callouts to Bills of Materials for variations on a standard theme.

The Configuration Drawing is a BOM (see section on BOMs). As such, it calls out everything shipped to the customer. Items to be included in the Configuration Drawing include:

- The unit itself (perhaps with dash numbers as above)
- Manual (with version number)
- Software disk
- Paper warranty notice
- FCC notice

Thus, starting with the Configuration Drawing, anyone can follow the "pointer trail" of BOMs and parts/drawings to figure out how to buy everything needed to make a unit, and then how to put it together.

Bills of Materials

A Bill of Material (BOM) lists *every* part needed for a subassembly.

The Drawing System really has only three sorts of drawings: BOMs, drawings for piece parts, and other engineering documentation. A piece part drawing is just like a part: it is something we build or buy and incorporate into a subassembly. As such, every piece part drawing is called out on a BOM, as is every piece part we purchase (like a 74HCT74). The part number of a piece part made from a drawing is just the drawing number itself. So, if drawing #1122 shows how to mill the product's baseplate, calling out part #1122 refers to this part.

"Other engineering documentation" refers to schematics, test procedures, modification drawings, ROM/PAL drawings, and assembly drawings (pictorial representations of how to put a unit together). None of these call out parts to buy, and therefore are always referenced on any BOM with a quantity of 0.

A piece part drawing can never refer to other parts; it is just one "thingy." A BOM always refers to other parts, and is therefore a collection of parts.

One BOM might call out another BOM. For example, the product A top-level BOM might call out parts (like the unit's box), drawings (like the baseplate), and a number of other BOMs (one per circuit board). In other words, one BOM can call out another as a part (i.e., a subassembly).

Though all BOMs have conventional four-digit drawing numbers, everything that refers to a BOM does so by appending a "dash number." That is, BOM #1234 is never called out on some higher-level drawing as "#1234"; rather, it would be "#1234-1" or "#1234-2", etc.

The dash number has two functions. First, it identifies the called out item as yet another subassembly. Any time you see a number with the dash number like this, you know that item is a subassembly.

The second reason is more important. The dash numbers let one drawing refer to several variations on a design. For example, if the BOM for the "Option A Memory Board" is drawing #1000, then #1000-1 might refer to 128k RAM and #1000-2 to 1 Mb RAM. The design is the same, so we might as well use the same drawings. The configuration is just a little different; one drawing can easily call out both configurations.

A good way to view the drawing system is as a matrix of pointers.

The Top Level Configuration Drawing (which is really a BOM) calls out subassemblies by referring to each with a drawing number with a dash suffix—a sort of pointer. Each subassembly contains pointers to parts or more levels of indirection to further BOMs. This makes it easy to share drawings between projects; you just have to monkey with the pointers. The dash numbers insure that every configuration of a project is documented, not just the overall PC layout.

BOM Format

BOMs are never "pictures" of anything—they are always just Bills of Materials (i.e., parts lists). The parts list includes every part needed to build that subassembly. Some of the parts might refer to further subassemblies.

The parts list of the BOM has the following fields:

- Item number (starting at 1 and working up)
- Quantity used, by dash number
- Part (or drawing) number
- Description
- Reference (i.e., U number or whatever)

Here is an example of a BOM #1000, with three dash number options. This is a portion of a memory option board BOM with several different memory configurations:

Item	Qty			Part #	Description	Ref
	-1	-2	-3			
1	#1000-1				OPTION board 256k	
2		#1000-2			OPTION board 1 mb	
3			#1000-3		OPTION board 4 mb	
4				#1892	OPTION ass'y	
5				#1234	OPTION schematic	
6				#1111	Test Procedure	
7	1	1	1	#1221	OPTION PCB	
8	8	8	8	Ap1123	32 pin socket	U1-8
9	1	1	1	74F373	IC	U10
10	8			62256	Static RAM	U1-8
11		8		621128	Static RAM	U1-8
12			2	624000	Static RAM	U1-2
13	2			APC3322	Jumper	J1,2
14		2	2	APC3322	Jumper	J3,4

First, note that each of the three BOM types (i.e., dash numbers) is listed at the beginning of the parts list. A column is assigned to each dash number; the quantities needed for a particular dash number are in this column. That is, there is a "quantity" column for each BOM type.

The first three entries, one per dash number, simply itemize what each dash number is. The quantity must be zero.

Each dash number column contains all quantity information to make that particular variation of the BOM.

Next, notice that drawing "#1892" is called out with a quantity of 0. Drawing #1892 shows how the parts are stuffed into the board, and is essential to production. However, it cannot call parts that must be bought, so it always has a quantity of 0.

The schematic and test procedure are listed, even though these are not really needed to build the unit. This is how all non-production engineering documents are linked into the system. All schematics, test procedures, and other engineering documentation that we want to preserve should be listed, but the quantity column should show 0. Notice also that a drawing number is assigned even to the test procedure. This insures that the test procedure is linked into the system and maintained properly.

The first column is the "item number." One number is assigned to each part, starting from 1 and working up. This is used where a mechanical drawing points out an item; in this case the item number would be in a circle, with an arrow pointing to the part on the drawing. It forms a cross reference between the pictorial stuffing drawing and the parts list. In most cases most item numbers will not have a corresponding circle on the drawing.

All jumpers that are inserted in the board are listed along with how they should be inserted (by the reference designator). This is the only documentation about board jumpering we need to generate.

Note that no modifications to the PCBs are listed. PC board modifications are to be listed on a separate "Mod" drawing, which is also referenced with a quantity of zero on the BOM.

ROMs and PALs

Every ROM and PAL used in a unit will be called out by two entries in the parts list columns of the PC board BOM. The first entry calls out the device part number (like GAL22V10) and associated data so purchasing can buy the part. The second entry, which must follow right after the first, calls out a ROM or PAL BOM.

The ROM or PAL BOM will be called out with quantity of 0. This procedure really violates the definition of the drawing system, but it drastically reduces the number of drawings needed by production to build a unit.

On the PC board BOM, the callout for a ROM or PAL will look like:

Item	Qty	Part #	Description	Ref
1	1	GAL22V10	PAL	U19
2	0	#1234-1	(MASTERS\PRODA\M-U19.PDS)	B9

Thus, the first entry tells us what to buy and where to put it; the second refers to engineering documentation and the current checksum. For a ROM, list the version number instead of the checksum. The description field for the part must also include the ROM or PAL's file name in parentheses, with directory on the lab computer.

ROMs, PALs, and SLD will be defined via BOMs, since these elements are really composed of potentially numerous sets of documentation. The ROM/PAL/SLD drawing will form the basic linkage to all source code files used in their creation.

The primary component of a PAL/ROM drawing is of course the device itself. Other rows will list the files needed to build the ROM or PAL.

Where two ROMs are derived from one set of code (like EVEN and ODD ROMs), these will both be on the same drawing.

An example ROM follows:

Item	Qty	Part #	Description	Ref
	-1			
	1	1234-1	64180 P-bd ROM	U9
1		27256-10	EPROM, 100 nsec	
2		PRODA.MAK—make file	proda\code	

Note that in this part list the EPROM itself is called out by conventional part number, but the quantity is 0 (since a quantity was called out on the PC board BOM that referenced this drawing).

A ROM, PAL, or SLD drawing calls out the ingredients of the device. In this case, the software's MAKE is listed so there's a reference from the hardware design to the firmware configuration.

If other engineering documentation exists, it should be referred to as well. This could include code descriptions, etc.

The last column contains the directory where these things are stored on the network drive.

The goal of including all of this information is to form one repository which includes pointers to all important parts of the component.

ROM and PAL File Names

All PALs and ROMs will have filenames defined by the conventions outlined here.

PALs are named: <board>-U<U number>.J<checksum>
ROMs are named: <board>-U<U number>.V<version>

Thus, you can tell a ROM from a PAL from the extension, whose first character is a V for a ROM or a J for a PAL.

Legal <board> names are: (limited to one character)

M - main board
P - option A board
T - option B board

Examples:

M-U10.JAB	main board, U10, checksum=AB
M-U1.J12	main board, U1, checksum=12

Engineering Change Orders (ECOs)

ECOs will be issued as required, in a timely fashion to insure all manufacturing and engineering needs are satisfied.

Every ECO is assigned against a drawing, not against a problem. You may have to issue several ECOs for one problem, if the change affects more than one drawing.

The reason for issuing perhaps several ECOs (one per drawing) is twofold. First, production builds units from drawings. They should not have to cross reference to find how to handle drawings. Secondly, engineering modifies drawings one at a time. All of the information needed to fix a drawing must be associated with the drawing in one place.

Each ECO will be attached to the affected drawing with a paperclip. The ECO stays attached only as long as the drawing remains incorrect. Thus, if you immediately fix the master (say, change the PAL checksum on the drawing), then the ECO will be attached to the newly Obsoleted Master, and filed in the Obsolete file.

If the ECO is not immediately incorporated into, say, a schematic, then the person issuing the ECO will pencil the change onto the Master drawing, so the schematic always reflects the way the unit is currently built.

In addition, if the ECO is not immediately incorporated into the drawing, the engineer issuing the ECO will mark the Master Drawing Book with the ECO and a brief description of the reason for the ECO, as follows:

Dwg #	Title	Revision	Rev Date	Filename
#3000	Prod A BOM	A	8-1-97	PRODA-ASSY
ECO: PRODA.A.3			Stabilize clk	PRODA.A.3
ECO: PRODA.A.1			Secure cables	PRODA.A.1

Note that the filename of the ECO is included in the Master Drawing Book.

When the ECO is incorporated into the drawing, remove the ECO annotation from the Master Drawing Book, as it is no longer applicable.

NEVER change a drawing without looking in the master repository to see if other ECOs are outstanding against the drawing.

Every change gets an ECO, even if the change is immediately incorporated into a drawing. In this case, follow the procedure for obsoleting a drawing. This provides a paper audit trail of changes, so we can see why a change was made, and what the change was.

Every ECO will result in incrementing the version numbers of all affected drawings. This includes the Configuration drawing as well. To keep things simple, you do not have to issue an ECO to increment the Configuration version number. We do want this incremented, though, so we can track revision levels of the products. Add a line to the Master Drawing Book listing the reason for the change and the new revision level of the Configuration, as well as a list of affected drawings. This forms back pointers to old drawings and versions. Though we remove old ECO history from our drawings, never remove it from the Configuration drawing's Master Drawing Book entry, as this will show the product's history.

The Master Drawing Book entry for an ECO'd Configuration drawing will look like:

Dwg #	Revision	Rev date	Title	Filename
#0600	A	8-1-97	Prod A Configuration	PRODA-ASSY
	B	8-2-97	Mod clock circuit to be more stable (1000-1, 1234 modified)	
	C	8-3-97	Secure cables better	

Sometimes a proposed ECO may not be acceptable to production. For example, a proposed mod may be better routed to different chip pins. Therefore, the engineer making an ECO must consult with production before releasing the ECO. (This avoids a formal (and slow) system of controlled ECO circulation.)

A decision must be made as to how critical the ECO is to production. The engineer issuing the ECO is authorized to shut down production, if necessary, to have the ECO incorporated in units currently being built.

Thus, to issue an ECO:

- Fill out the ECO form, one per drawing, and distribute it to production and all affected engineers.
- If you don't immediately fix the drawing, clip it to the affected drawing and mark the Master Drawing Book as described.
- If necessary, pencil the changes onto the Master drawing.
- Increment the Configuration Drawing version number immediately. Add a line to the Master Drawing Book after the Configuration drawing entry describing the reason for the change, and listing the affected drawings.
- If the change is a mod, consult with production on the proposed routing of the mod.
- If the change is critical, instruct production to incorporate it into current work-in-progress.
- Remember that most likely several drawings will be affected: a new mod will affect the schematic and the BOM that shows the mod list.

To incorporate an ECO into a drawing:

- Make whatever changes are needed to incorporate ALL ECOs clipped to that drawing.
- Revise the version letter upwards.
- Generate a new Master drawing, and Obsolete the old Master.
- Delete the ECO file from the network drive.
- Revise the Version letter on the Configuration drawing.

Responsibilities

The engineer making a change is responsible to insure that change is propagated into the drawing system, and that the information is disseminated to all parties. He/she is responsible for filing the drawings, removing and refiling obsoleted drawings, stamping MASTER or OBSOLETE, etc.

The engineer making the change must update production's master ROM/PAL computer with current programming files, and the drawings with checksums and versions as appropriate. The engineer must immediately also update the network drive, and pass out ECOs.

Nothing in this precludes the use of clerical staff to help. However, final responsibility for correctness lies with the engineer making changes.

The Master Drawing Book does contains information about "Specials" we've produced. The manufacturing technician is responsible to insure that all appropriate information is saved both in this Book and in the unit's folder.

The production lab MUST maintain an accurate, neat book of CURRENT BOMs, to insure the units are built properly. Every change will result in an ECO; the lab must file that promptly.

Index

Access, nonintrusive, 136–37
Addresses
  logical, 94
  translating, 96
ALE (Address Latch Enable), 117
Analysis, post mortem, 194–95
Analyzers
  logic, 158
  performance, 79–82
ASICs (application-specific integrated circuits), 76, 109, 142, 154
Assembly
  formatting, 218–19
  language, 61–64
Assumptions, 172–74
Audit, weekly, 187
Author's role defined, 17

Bad code, identify, 30
Banking, 93–97
  hardware issues, 94–96
  logical to physical, 94
  software, 96–97
BDM (Back-ground Debug Mode) and JTAG (Joint Test Access Group) hardware, 143–44
BDMs (Back-ground Debug Modes), 142–45, 162, 184
  debugger, 144
Bit banging software, UART, 44
BOMs (Bills of Materials), 224, 229–30
Bond-out chips, 140
Book, Master Drawing, 226–27
Boss management, 190–92
Breakpoints
  complex, 138
  hardware, 40, 138
  problems, 69–71
Bug measurements, three big reasons for, 27–28
Bug rates
  measure one's, 27–30
  identify bad code, 30
  stop, look, listen, 28–30

C
  formatting, 217–18
  language, 61–64
Capital equipment justification, 155
Challenger explosion, 1, 192
Chips
  bond-out, 140
  FIFO, 60–61
CIMM (Capability Immaturity Model), 9–10
Clip leads, 171, 177
Clock-shaping logic, 117
Clocks, 115–17
CMM (Capability Maturity Model), 8–33
  achieving schedule and cost goals, 10
  being wary of, 12
  five levels of software maturity, 9
CMOS (complementary metal-oxide semiconductors), 112, 151
  gate, 113
  logic, 114
  voltage levels, 116
COCOMO (Constructive Cost Model)
  data, 36–37
  metric, 41
  model, 37
Code
  break down by features, 47
  complexity grows much faster than program size, 82–83
  cost of inspecting, 22
  how fast one generates embedded, 32
  Inspections, 133
  startup, 207–8
  writing polled, 54–55
Code Inspections
  process, 18–22
    follow-up, 20
    inspection meeting, 19–20

237

238 THE ART OF DESIGNING EMBEDDED SYSTEMS

Code Inspections *(continued)*
   miscellaneous points, 20–22
   overview, 18–19
   planning, 18
   preparation, 19
   rework, 20
   teams, 17–18
Code production rates, measuring one's, 31–32
Codes, create, compile, and test, 90
Coding conventions, 216–19
   assembly formatting, 218–19
   C formatting, 217–18
   general, 216
   spacing and indentation, 216–17
COGS (cost of goods), NRE versus, 42–43
Comments, 215–16
Compiler vendors, 62–63
Compilers, 220
Complex breakpoints, 138
Complexity does not scale linearly with size, 35
Computers
   timing is critical in, 174
   tools, 219
Configuration Drawings, 227–28
Connections, reliable, 158–59
Cost
   of inspecting code, 22
   payroll as fixed, 153
CPUs (central processing units), 41, 54–56, 61, 64–65, 77, 118, 120, 185
   partitioning with, 40–44
   simplifying software through multiple, 43–44
Cubicles, working in, 25–26

Data
   COCOMO (Constructive Cost Model), 36–37
   collecting, 28
   presenting, 28
Data-destroying event, 14
Data sheets
   notes of, 118
   read, 118

Datacomm problems, 70
Debug bit, 80
Debuggers
   BDM (Back-ground Debug Mode), 144
   BDM-like, 59
   features, 135–39
   JTAG (Joint Test Access Group), 144
Debugging, 220–21
   basic philosophy of, 165
   easy ISR, 71–72
   INT/INTA cycles, 64–66
   scope, 178–83
   source-level, 135–36
   tool vendors, 159–61
   traces change philosophy of, 70
Debugging port, virtual, 180
Debugging resources, add, 161–62
Degrees of higher learning, 197–201
Delayed sweep, 180–82
Design process, and human nature, 49
Designing products, improving process of, 193
Designs
   correct, 112
   debuggable, 109–11
   top-down, 37
   watchdog, 124
Developers, ideal prototype, 108
Development, disciplined, 5–34
Devices
   manual testing of, 90
   mastering portions of, 89–90
   overheating, 176
   refreshing, 103
Diagnostics, RAM, 98–104
Directory structure, 204–5
Discipline, engineering is very diverse, 200
Disciplined development, 5–34
DMA (direct memory access), 90, 161
Documentation, 171–72
DRAMs (dynamic random-access memories), 102–3
Drawing Book, Master, 226–27
Drawing system, simple, 223–35
   BOMs (Bills of Materials), 228–30
   Configuration Drawing, 227–28

drawings and drawing storage, 224–26
ECOs (Engineering Change Orders),
    232–34
Master Drawing Book, 226–27
responsibilities, 234–35
ROM and PAL file names, 232
ROMs and PALs, 230–32
Drivers, hacking peripheral, 87–90

ECOs (Engineering Change Orders),
    226, 232–34
Electrical noise, 102
Embedded code, how fast one generates,
    32
Emulation RAM, 137–38
Emulators, 139–42
    downsides of, 141–42
    ROM, 112, 146
Encapsulation, partitioning with, 38–40
Environment, creating quiet work, 22–27
EOI (end of interrupt), 66
EPROMs (erasable programmable read-
    only memories), 121–22, 129
Equipment
    capital, 155
    leasing, 157
    soldering, 170
Estimate, learn to, 174–78
Estimation, one of engineering's most
    important tools, 77
Event, data-destroying, 14
Experience, 77–78
    practical, 74
    value of, 6

Feature matrix, 46–47
Features
    break down codes by, 47
    partitioning by, 45–58
Feedback loop
    close, 78
    managing, 192–96
FIFO (first-in, first-out) chips, 60–61
File names, ROM and PAL, 232
Files
    make, 207
    project, 207
    version, 205–6

Filters, event triggers and, 137
Firmware
    costs of, 7
    development incrementally, 48–50
    estimate performance of, 174–75
    banking, 93–97
    curse of Malloc( ), 92–93
    hacking peripheral drivers, 87–90
    notes on software prototyping,
        104–8
    predicting ROM requirements,
        97–98
    RAM diagnostic, 98–104
    selecting stack size, 90–92
    testing, 48
Firmware standard, Code Inspections,
    21
Firmware standards manual, 203–21
    coding conventions, 216–19
        assembly formatting, 218–19
        C formatting, 217–18
        general, 216
    comments, 215–16
    functions, 214
    institute, 15–16
    ISRs (Interrupt Service Routines),
        214–15
    modules, 209–12
        general, 209
        names, 212
        templates, 209–12
    projects, 204–9
        directory structure, 204–5
        heap issues, 208–9
        make files, 207
        project files, 207
        stack issues, 208–9
        startup code, 207–8
        version file, 205–6
    scope, 203–4
    tools, 219–21
        compilers, 220
        computers, 219
        debugging, 220–21
    variables, 212–13
        global, 213
        names, 212–13
        portability, 213

Formatting, assembly, 218–19
FPGAs (field-programmable gate arrays), 129
Functions, 214
　most of bugs will be in few, 30
　and reentrants, 67
　using to do one thing, 59

Gate, CMOS, 113
Glitches, diagnose all, 174
Global variables, 68, 213
Globals, 38
Grounders, using clip leads as, 177
Guesstimating, 75–76

Hacking peripheral drivers, 87–90
Handlers, keep short, 58
Hardware
　breakpoints, 40, 138
　is moving away from conventional prototypes, 105
　issues, 59–61, 94–96
　　changing PCBs (printed circuit boards), 128–30
　　clocks, 115–17
　　debuggable designs, 109–11
　　making PCBs (printed circuit boards), 126–28
　　planning, 130–31
　　reset, 117–19
　　resistors, 111–13
　　small CPUs, 119–23
　　unused inputs, 114–15
　　watchdog timers, 123–26
Hardware design, let software drive, 40
Heap issues, 208–9
Heat, being on lookout for excessive, 176
　See also Overheating
Human nature and design process, 49

ICEs (In-Circuit Emulators), 139, 184
ICs (integrated circuits)
　See also Chips
　software, 74
Idle loops, 81–82
Idle time, 81
Impossible, conquer, 50–51
Inheritance, 38

Inputs
　unused, 114–15
　　leave unconnected when building prototypes, 115
Inspection team, keep management off, 17
Inspections, use Code, 16–22
INT/INTA cycles, debugging, 64–66
Integration, 48
Interrupt map, lay out, 57–58
Interruptions from work, 25
Interrupts; *See also* ISRs (interrupt service routines), 54–64
　C or assembly languages, 61–64
　design guidelines, 57–59
　finding missing, 66–67
　hardware issues, 59–61
　from internal peripherals, 64
　latency of, 80
　vectoring, 55–57
INTR signal, generation of, 60
ISRs (interrupt service routines), 40, 54–55, 57, 214–15
　approximate complexity of, 58
　cardinal rule of, 58
　easy debugging, 71–72
　keeping simple, 59
　using complex data structures in, 63

JTAG (Joint Test Access Group), 143, 162
　and BDM (Back-ground Debug) hardware, 143–44
　debuggers, 144

Keyboard, seduction of, 5
Knives, X-Acto, 129–30, 152
Knowledge is power, 91

Languages
　assembly, 61–64
　C, 61–64
　CMSP, 63
　writing shells of drivers in selected, 89
LCDs (liquid crystal displays), 166
Leads, clip, 171
Leasing most attractive way to get equipment, 157

Index **241**

LEDs (light-emitting diodes), 121, 178
LOC (lines of code), 46, 97–98
Logic
  analyzers, 158
  clock-shaping, 117
  CMOS, 114
Logical address, 94
Loops, idle, 81–82
LS (large-scale) technology, 151

Make files, 207
Malloc( ), curse of, 92–93
Management
  boss, 190–92
  defined, 190
  engineering, 194
  keep off inspection team, 17
  of oneself, 187–90
Managers, Peopleware argument with, 27
Manual, institute firmware standards, 15–16
Manual testing of devices, 90
Map, lay out interrupt, 57–58
Market, Time To, 154, 199
Mars Pathfinder spacecraft, 173–74
Master Drawing Book, 226–27
Matrix, feature, 46–47
Media, will unreadable tomorrow, 15
Memory
  OTP (One-Time Programmable) program, 121–22
  problems, 99
Microcontrollers, 123, 140
Midrange processors, 123
Models of products, virtual, 107
Moderator defined, 17
Module design, something profound about, 40
Module names, 212
Modules
  defined, 209
  most of bugs will be in few, 30
Money, time costs, 155
Monitors
  ROM, 145–46
  watchdog, 125
Myths, nonintrusive, 159–61

Names, ROM and PAL file, 232
Network computing lets users share data, 73
NMIs (non-maskable interrupts), 112–13, 124
  avoiding, 69
  reoccurs at any time, 70
Noise
  electrical, 102
  issues, 101–4
  when digital systems are most susceptible to, 102
Nonintrusive access, 136–37
Nonintrusive myths, 159–61
NRE costs (nonrecurring engineering costs), 42–43
NRE versus COGS, 42–43
Numbers, interpreting raw, 28

OOPs (object-oriented programs), 37, 84
Operating systems give tools to manage resources, 84
Oscilloscopes; See also Scopes; Scoping tricks, 147–52
  favorite software debugging tools, 147
  and timing, 149
  triggering signals, 150
OTP (One-Time Programmable) program memory, 121–22
Output bits for debugging purposes, 79
Overheating devices, 176
Overlay RAM, 137–38

PAL file names, ROM and, 232
PAL (programmable array logic), 121, 129, 167–69
  and ROMs, 230–32
Partitioning, 37–48
  with CPUs, 40–44
  with encapsulation, 38–40
  by features, 45–48
Parts, surface-mount, 129
Pattern sensitivity, 101
Payroll as fixed cost, 153
PCBs (printed circuit boards), 101–2, 110, 126–28
  changing, 128–30
  defects, 177

PCMCIA (Portable Computer Memory Card International Association), 159
People musings, 187–201
  boss management, 190–92
  degrees, 197–201
  managing feedback loop, 192–96
  managing oneself and others, 187–90
    bug management, 188–89
    critical paths, 190
    firmware standards, 188
    tools, 189
    tracking development rates, 189
    version control system, 188
    work environment, 189–90
Peopleware *(DeMarco and Lister)*, 22
Peopleware argument with managers, 27
Performance
  analyzer, 79–82
  guesstimating, 72–79
  measuring, 72–82
Peripherals
  drivers
    fraught with risks and unknowns, 87
    hacking, 87–90
    incredibly complex, 65
    interrupts from internal, 64
Personal Software Process, 33
Physical space, 94
Plan ahead, 176
Planning, 130–31
PLDs, 121, 128–29
Polled code, writing, 54–55
Polymorphism, 38
Ports
  using serial, 88
  virtual debugging, 180
Post mortem
  analysis, 194–95
Probes, take care of oscilloscope, 150
Problems
  breakpoint, 69–71
  datacomm, 70
  expect, 134
  reentrancy, 67–69

Problems, solving, 2, 12
Production rates, measuring one's code, 31–32
Productivity, 35
Products
  customers and views of, 45
  improving process of designing, 193
  quality of, 8
  virtual models of, 107
Products, shipping quality, 47
Profession, worry for future of engineering, 199
Professionals creating software, 6
Program size, code complexity grows much faster than, 82–83
Programming languages; *See* Languages
Programming, structured, 37
Programs, stop writing big, 35–51
  COCOMO (Constructive Cost Model) data, 36–37
  conquer impossible, 50–51
  develop firmware incrementally, 48–50
  partitioning, 37–48
Project files, 207
Prototype code, writing in Visual Basic, 107
Prototype developers, ideal, 108
Prototypes, 106, 134
  hardware is moving away from conventional, 105
  of system's software, 105
Prototyping, notes on software, 104–8
Pull-down resistors, 112–13, 160
Pull-up resistors, 113, 160

Quality
  is nice, 7–8
  of products, 8
Quality products, shipping on time, 47

RAM (random-access memory), 58, 99–103, 119, 185
  diagnostics, 98–104, 100–101
    inverting bits, 100–101
    noise issues, 101–4
  emulation, 137–38

Index **243**

overlay, 137–38
shadow, 138
Reader defined, 17
Real-time trace, 137
Recorder defined, 17
Reentrancy problems, 67–69
Refreshing devices, 103
Renting equipment, 156
Reset, 117–19
   glitches, 173–74
   time delay on, 118
Resistors, 111–13
   pull-down, 112–13, 160
   pull-up, 113, 160
Resources, operating systems give tools to manage, 84
Responsibilities, simple drawing system, 234–35
Results, define, 106
Rise and fall times, 117
RMAs (rate monotonic analysis) and schedulers, 83
ROM emulators, 112, 146
ROMs (read-only memories), 129, 185
   monitors, 145–46
   and PAL file names, 232
   and PALs, 230–32
   requirements, 97–98
RS-232, one of biggest headaches around, 179
RTOSs (real-time operating systems), 81–85, 96, 125, 194
   is context switcher, 83
   using, 85

SCC (Serial Communications Controller), Zilog, 183
Schedulers and RMAs, 83
Schedules, 190
   collapse of, 31
Schematics, 128
Scopes; *See also* Oscilloscopes
   debugging by, 178–83
   grounding, 152
   simple drawing system, 223–24
   tricks to effective uses, 180

Scoping tricks, 150–52
SCR latchup, 115
SCR (silicon controlled rectifier), 114
Sensitivity, pattern, 101
Serial ports, using, 88
Seven-step plan, 12–33
   buying and using VCS (Version Control System), 13–15
   constantly study software engineering, 32–33
   creating quiet work environment, 22–27
   instituting firmware standards manual, 15–16
   measuring one's
      bug rates, 27–30
      code production rates, 31–32
   using Code Inspections, 16–22
Shadow RAM, 138
Shorts, 175
Signals
   generation of INTR, 60
   triggering, 150
SMT (surface-mount technology), 129, 142, 152
Sockets, 129
Software
   debugging, 79
   drives hardware design, 40
   engineering, 32–33
   ICs, 74
   professionals creating, 6
   prototypes of system's, 105
   prototyping, 104–8
   simplifying through multiple CPUs, 43–44
   UART bit banging, 44
Software maturity, CMM defines five levels of, 9
Soldering
   equipment, 170
   inspecting, 177
Source debugger, 97
Source-level debugging, 135–36
Space, physical, 94
Spacecraft, Mars Pathfinder, 173–74
Spikes, timing, 119

Spreadsheets, 107
SRAM (static random-access memory), 119
Stack
  issues, 208–9
  size, 90–92
Stamping, time, 139
Startup code, 207–8
Stimulus, creating, 88
Structured programming, 37
SWAN (Smart, Works hard, Ambitious, and Nice) model, 200
Sweep, delayed, 180–82
Switches and embedded systems, 126
System
  bringing up new, 183–85
  total idle time of, 81
System status info, embedded systems and managing, 84
System's performance, tracking, 78
System's response, measuring, 88

Target processor, developing understanding of, 77
Teams, Code Inspections, 17–18
Technicians
  turned-engineers, 200
Technology, LS, 151
Templates, 209–12
Test equipment, never blindly trust, 173
Testing
  daily or weekly, 49
  everything, 173
  firmware, 48
  points, 109–11
  success requires determination to constantly, 49
Think, need to focus to, 26
Time
  costs money, 155
  idle, 81
  to market, 154, 199
  real, 53–85
    avoiding NMI (non-maskable interrupt), 69
    breakpoint problems, 69–71
    debugging INT/INTA cycles, 64–66
    easy ISR debugging, 71–72
    finding missing interrupts, 66–67
    interrupts, 54–64
    measuring performance, 72–82
    reentrancy problems, 67–69
    RTOS, 82–85
  stamping, 139
Timers, watchdog, 123–26
Timing
  details, 161
  is critical in computers, 174
  and oscilloscopes, 149
  spikes, 119
Tool vendors, debugging, 159–61
Tool woes, 157–63
  add debugging resources, 161–62
  nonintrusive myths, 159–61
  reliable connections, 158–59
  ROM burnout, 162–63
Tools, 134–52
  checkpointing, 15
  CMMs (Capability Maturity Models) are, 12
  compilers, 220
  computers, 219
  debugging, 220–21
  quest to obtain right, 156
  scope complements, 178
  troubleshooting, 133–63
    BDMs (Back-ground Debug Modes), 142–45
    cost of, 152–57
    emulators, 139–42
    fancy, 152–57
    oscilloscopes, 147–52
    ROM emulators, 146
    ROM monitors, 145–46
    tool woes, 157–63
  use all, 177
Tools to manage resources, operating systems give, 84
Top-down design, 37
TQFP, 158
Traces, 80
  change philosophy of debugging, 70
  real-time, 137
Trigger levels, 181
Triggering signals, 150
Triggers, event, 137

Troubleshooters, best, 176
Troubleshooting, 165–85
  bringing up new system, 183–85
  scope debugging, 178–83
  sequence, 166–69
    fix bug, 169
    generate experiment to test hypothesis, 167–69
    generate hypothesis, 167
    observe behavior to find apparent bug, 166
    observe collateral behavior, 166–67
    round up usual suspects, 167
  speed up by slowing down, 169–78
    assumptions, 172–74
    documentation, 171–72
    learn to estimate, 174–78
  tools, 133–63
    BDMs (Back-ground Debug Modes), 142–45
    emulators, 139–42
    oscilloscopes, 147–52
    ROM emulators, 146
    ROM monitors, 145–46
    scoping tricks, 150–52
Trust between workers and management, 191
TTL (transistor-transistor logic), 115–16

UARTs (universal asynchronous receiver-transmitters), 54, 57, 66, 121, 183
  bit banging software, 44
Understanding, good measures promote, 28

Variables, 212–13
  avoiding global, 68
  declared as static, 68
  global, 213
  names, 212–13
  portability, 213

VCS (Version Control System), 13–15, 205
Vectoring, 55–57
Vendors, compiler, 62–63
Version file, 205–6
Virtual corporation, 157
Virtual debugging port, 180
Virtual instruments, 106
Virtual models of products, 107
Visual Basic, writing prototype code in, 107

Watchdog
  design, 124
  monitors, 125
  timers, 123–26
WDTs (watchdog timers), 123–26
  and safety issues, 125
Weekly audit, 187
Work
  environment, 22–27
  interruptions from, 25
Workers and management, trust between, 191
Writing
  few engineering programs focus on, 199
  polled code, 54–55
Writing big programs, stop, 35–51
  COCOMO (Constructive Cost Model) data, 36–37
  conquer impossible, 50–51
  develop firmware incrementally, 48–50
  partitioning, 37–48

X-Acto knives, 129–30, 152

Z80 processors, 66
Z180 processors, 66, 117–18
Zilog SCC (Serial Communications Controller), 183